BRAVE RIFLES
The Theology of War

by

Bradford Smith

Published by

Olivia Kimbrell Press™

Copyright Notice

Brave Rifles: The Theology of War

First edition. Copyright © 2017 by Bradford Smith. All rights reserved. No part of this publication may be reproduced or transmitted in any form or by any means—electronic, mechanical, photocopying, or recording—without express written permission of the author. The only exception is brief quotations in printed or broadcasted critical articles and reviews. This book is a work of nonfiction. Specific names and places of individuals and towns may have been changed in order to protect the privacy of those involved or where required by law in the case of minor children.

PUBLISHED BY: Olivia Kimbrell Press™*, P.O. Box 470, Fort Knox, KY 40121-0470

The Olivia Kimbrell Press™ colophon and open book logo are trademarks of Olivia Kimbrell Press™. *Olivia Kimbrell Press™ is a publisher offering true to life, meaningful fiction from a Christian worldview intended to uplift the heart and engage the mind.

Some scripture quotations courtesy of the King James Version (KJV) of the Holy Bible. Some scripture quotations taken from the Holman Christian Standard Bible® (HCSB), Copyright© 1999, 2000, 2002, 2003, 2009 by Holman Bible Publishers. Used by permission. All rights reserved. Some scripture quotations courtesy of the New King James Version (NKJV) of the Holy Bible, Copyright© 1979, 1980, 1982 by Thomas-Nelson, Inc. Used by permission. All rights reserved.

Original Cover Art by Amanda Smith (www.amandagailstudio.com)

Library Cataloging Data

Names: Smith, Bradford (Smith Bradford) 1973-

Title: Brave Rifles; The Theology of War / Bradford Smith

448 p. 6 in. × 9 in. (15.24 cm × 22.86 cm)

Description: digital eBook edition | Print on Demand edition | Trade paperback edition | Kentucky: Olivia Kimbrell Press™, 2017.

Summary: The warrior before God from the Bible to the battlefield.

Identifiers: ePCN: 2017959217 | ISBN-13: 978-1-68190-110-7 (ebk.) | 978-1-68190-108-4 (POD) | 978-1-68190-109-1 (trade)

1. military theology God 2. soldier prayer war 3. relationship suicide 4. justified warfare honor 5. trauma atrocity PTSD 6. Jesus combat love 7. US Army fight win

BRAVE RIFLES
The Theology of War

by

Bradford Smith

If only one...

Table of Contents

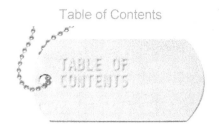

BRAVE RIFLES: THE THEOLOGY OF WAR

INTRODUCTION   1

   Preparatory Fires   3
  Chapter 1   **1**
    1.   Approaching Jericho   1
  Chapter 2   **7**
    2.   Context of the Examination   7
  Chapter 3   **17**
    3.   First Things, Hardest Things   17

SECTION 1: WAR AND THE BODY   35

  Chapter 4   **37**
    4.   The Source of War   40
  Chapter 5   **55**
    5.   Road to War   55
  Chapter 6   **69**
    6.   War and Sovereignty   69
  Chapter 7   **81**
    7.   Lambs and Lions   81
  Chapter 8   **95**
    8.   Centurions and Soldiers   95
  Chapter 9   **109**
    9.   Saved Rounds   109

SECTION 2: WAR AND THE MIND   117

  Chapter 10   **119**
    10.   The Utility of War   124
  Chapter 11   **137**
    11.   The Mind of the Warrior   137
  Chapter 12   **145**

12.	The Struggle of War	145
Chapter 13		**157**
13.	Killing the Mind	157
Chapter 14		**171**
14.	Resolution	171

SECTION 3: WAR AND THE HEART 185

Chapter 15		**187**
15.	Hearts on Display	189
Chapter 16		**195**
16.	The Theoretical Heart	195
Chapter 17		**209**
17.	The Heart in Practice	209
Chapter 18		**219**
18.	Heart of a Leader	219
Chapter 19		**229**
19.	Consecrated Hearts	229

SECTION 4: WAR AND THE SOUL 233

Chapter 20		**235**
20.	The Soul and the Law	238
Chapter 21		**247**
21.	The Soul of a Warrior	247
Chapter 22		**257**
22.	War and the Soul	257

SECTION 5: WAR AND THE NATIONS 271

Chapter 23		**273**
23.	God and the Nations	275
Chapter 24		**291**
24.	Problem of a Godless Army	291
Chapter 25		**299**
25.	Danger of a Godless Army	299
Chapter 26		**307**
26.	Sex in a Godless Army	307
Chapter 27		**321**
27.	Affliction of a Godless Army	321

Final Fires	**341**
BIBLIOGRAPHY	**345**

DEEP DIVE COMPANION BIBLE STUDY — 353
ABOUT THIS STUDY GUIDE — 355
___The Deep Dive Companion — 355
___Study Guide Methodology — 359

INTRODUCTION STUDY GUIDE FOR BRAVE RIFLES — 361
Preparatory Fires — 363
Study Guide for Chapter 1 — 365
Approaching Jericho — 365
Study Guide for Chapter 2 — 367
Context of the Examination — 367
Study Guide for Chapter 3 — 369
First Things, Hardest Things (part 1) — 369
Study Guide for Chapter 3 — 371
First Things, Hardest Things (part 2) — 371

STUDY GUIDE SECTION 1: WAR AND THE BODY — 373
Study Guide for Chapter 4 — 375
The Source of War — 375
Study Guide for Chapter 5 — 377
Road to War — 377
Study Guide for Chapter 6 — 379
War and Sovereignty — 379
Study Guide for Chapter 7 — 381
Lambs and Lions — 381
Study Guide for Chapter 8 — 383
Centurions and Soldiers — 383
Study Guide for Chapter 9 — 385
Saved Rounds — 385

STUDY GUIDE SECTION 2: WAR AND THE MIND — 387
Study Guide for Chapter 10 — 389
The Utility of War — 389
Study Guide for Chapter 11 — 391
The Mind of the Warrior — 391
Study Guide for Chapter 12 — 393
The Struggle of War — 393
Study Guide for Chapter 13 — 395
Killing the Mind — 395
Study Guide for Chapter 14 — 397

Resolution ... 397

STUDY GUIDE SECTION 3: WAR AND THE HEART — 399

Study Guide for Chapter 15 — 401
  Hearts on Display — 401
Study Guide for Chapter 16 — 403
  The Theoretical Heart — 403
Study Guide for Chapter 17 — 405
  The Heart in Practice — 405
Study Guide for Chapter 18 — 407
  Heart of a Leader — 407
Study Guide for Chapter 19 — 409
  Consecrated Hearts — 409

STUDY GUIDE SECTION 4: WAR AND THE SOUL — 411

Study Guide for Chapter 20 — 413
  The Soul and the Law — 413
Study Guide for Chapter 21 — 415
  The Soul of a Warrior — 415
Study Guide for Chapter 22 — 417
  War and the Soul — 417

STUDY GUIDE SECTION 5: WAR AND THE NATIONS — 419

Study Guide for Chapter 23 — 421
  God and the Nations — 421

STUDY GUIDE CONCLUSION — 423

PERSONAL NOTE — 430

ACKNOWLEDGEMENTS — 431

ABOUT THE AUTHOR — 432

Introduction

Bradford Smith

Brave Rifles: The Theology of War

Introduction

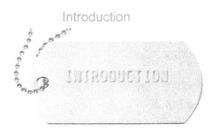

The society that separates its scholars from its warriors will have its thinking done by cowards and its fighting done by fools…Thucydides 460-400BC

Preparatory Fires

Saturday, February 27th

"You're up," Mike confirmed. "We need you."

My heart leaped in my chest and butterflies flooded my gut right there by the milk section of the local Wal-Mart.

I hadn't been to the show in a few years and even worse, I had been left behind to command the rear detachment on a recent deployment, a fate far worse than death as anyone who has endured this indignity will attest. Few things humble a soldier like watching your brothers-in-arms head to battle without you. I remember telling my boss, a great friend of mine and an officer for whom I have tremendous respect, that I would never forgive him for leaving me behind. Unfortunately, it made the most sense.

Nevertheless, I stewed in the rear for ten excruciating months, concerning myself with trivialities, inspections, unit statistics and measurements, all manner of garrison whitewashing. Meanwhile, my comrades covered themselves in glory on the battlefield, taking the fight to the wicked men and entities who stand opposed to all that we cherish. They performed magnificently.

They didn't lose a single soldier in combat though we lost six on rear detachment. Pride swelled my heart on the day my brothers stepped off the plane, any bitterness quickly forgotten amidst the joyous reunification of family and friends. I hadn't realized how my heart yearned to be there alongside and what a bitter pill it had been to swallow, staying behind.

When I got my orders sending me back to the unit, I quickly got on the horn and started gathering information. I called the executive officer, an old friend of mine, who let me know they were expecting me and might need me for a short deployment. Surprised, I told him I was more than available and would be honored but didn't think too much about it. Honestly, I wasn't expecting to get the nod.

So I was more than surprised when I ran into Mike and he confirmed.

"Where to?" I inquired.

"Iraq."

The subtle humor of two middle-aged men lingering near the milk section of Wal-Mart discussing a secret war against ISIS while trying to maintain control of our kids struck me. Not wanting to violate operational security any further, I didn't ask any more questions. I told him I intended to sign back into the unit in a week or so and I'd come get an operational lay-down.

Iraq. I hadn't been to Iraq since 2010 and frankly, never intended to return. As a matter of fact, I prayed in earnest that the fight would end such that my sons would never have to fight there. The good Lord was going to give me another chance to make good on this desire.

I hardly heard a thing as I finished off the grocery list Ami had sent with me. Memories flooded my brain, memories of days long gone, of battles fought and brothers lost. My mind immediately began to visualize what I'd be doing, who I'd be with, the steps and things I needed to accomplish to get there. I needed to get a read-on, an operational lay-down. I needed to get some new kit, get to the range. I needed to update powers of attorney and any other paperwork...and I needed to tell Ami.

That last part was always the hardest. My wife is the greatest trooper I know, yet for some reason, I always hesitated to tell her and I do mean always, perhaps searching for the "right" time, as if there were such a thing. One particular incident highlighted my ridiculous attempts to handle my wife with the kid gloves I seemed to think were necessary.

I had just returned from a stint in Iraq, maybe two months long—my unit deployed on frequent but short in duration trips, unlike the yearlong deployments that became customary across most of the Army. About a week after returning, I got word that we were dispatching a group back to Afghanistan to fight the Taliban. I had not been to Afghanistan since 2003 as fighting in Iraq consumed most of our unit since the invasion. I immediately began lobbying the headquarters to send my company.

I assembled my planners, a few big brains from the company, and we produced some quick charts, replete with multi-color bar graphs and timelines that showed why it made complete operational sense to send us. Chris, my senior warrant officer, advised caution, "A bird in the hand is worth two in the bush". We were offering to bear the burden up front in exchange for some slack at a later date. Chris turned out to be 100% correct, not that I'd ever tell him.

But, we got what we wanted. I would deploy my company to fight the Taliban on their turf, the Helmand province of southern Afghanistan. I never suspected that I'd spend the next three years in and out of that God forsaken desert. Leaving in two weeks, I knew time was already short. Now, how to tell Ami?

With a heavy heart, I drove home from work that day. I'd only been back a week. What would she say? How would she react? What about the girls? Would they cry? It was a sweaty July afternoon as I pulled up to our home. Ami, a serial reorganizer, had the garage door up and most of our junk strewn across the driveway. Instantly, she could tell something was wrong as I plodded up the driveway, my feet heavy.

I plopped down and just spat it out. "I'm deploying again."

"When?"

"Two weeks," I muttered, choking back tears. Looking up, "I'm sorry, baby."

"How long?"

"Two, maybe three months."

She hesitated for maybe a second and then, "Does this mean you'll be home for Christmas now?"

What was this? A glimmer of hope? "More than likely," I offered hesitantly.

"Then go knock it out and get back to us for Christmas."

My spirit soared! I couldn't believe what a trooper my wife was though I don't know why I doubted. She came into our marriage a young, naive twenty-something-year-old and then a year later, the war started. I spent the next decade coming and going while she held the fort down and raised our girls. I cannot imagine how difficult that must've been but she never complained. Her only demand was that when I was home, I was home.

Though I hated leaving her and the girls, coming home always made it worthwhile and almost worth going in the first place. It was the good-byes.

I dreaded the good-byes almost more than the deployment itself. The days just before leaving were always ferociously painful and absolutely excruciating. I just wanted to get there, get the clock punched, and get home.

I had a switch. I discovered it years before at the airport. I don't remember when it was or where I was going, but Ami and the girls drove me to the airport as I was flying commercial on this particular deployment. At the airport, all three of the girls were misbehaving, acting out in anticipation of me being gone. I struggled between not wanting to fuss at them in our last few moments together and wanting them to calm down. As I embraced Ami a man approached and interrupted. Though I was not in uniform, he must've gathered what he was witnessing.

"Excuse me, Sir. Are you a deploying soldier?" He asked.

"I am."

"I didn't want to interrupt, but just wanted to thank you for your service."

I gratefully accepted his thanks before turning back to my family. With a heavy heart, I bid Ami farewell, kissed the girls and gathered my things. Once I cleared security, I quickly occupied a seat behind a large sign that would conceal me a bit and I sat down and—don't judge me—cried, quietly, periodically peeking to ensure no one was watching. After a minute, I gathered myself, girded up my loins, and uttered an internal, "Okay, let's do this."

I flipped the switch. Good-byes complete; time for business. I could press start on the countdown clock and anticipate the first warm embrace of my beautiful wife and the joyous squeezes from my girls.

Now, though, I had to tell Ami once more and as I had not deployed in a few years, I wasn't quite sure what to expect. We had also since added the boys to the clan.

I hesitated for my usual day or two before I finally just blurted it out, "The unit's deploying me."

"When?"

"About the 1st of May."

"How long?"

"Maybe two months."

"Okay."

"That's it."

"Yes."

"It should be my last one."

"Okay."

"Well okay," I finally responded, not quite sure what to say. "I'll tell the people." I had become unfortunately infamous in my home for forgetting to tell the girls when I was leaving to go out of town. Vowing not to replicate the error, I set about informing the people.

"Daddy's going to fight the bad guys," I confided in my little guys. I had intermixed our nightly bedtime stories with frequent tales from the battlefield so they were well versed in the need and duty to fight the bad guys.

My girls received the news with the matter-of-fact nature that only a military family will ever understand. I knocked on my youngest daughter's bedroom door. "I have cancer and I'm deploying," I told her as she peeked out, wasting no time with sugar-coating—I had recently been diagnosed with a basal cell carcinoma on my head, mild skin cancer.

"When?"

"First of May for two months."

"You'll miss my birthday," she informed me. I hung my head in repentance.

"I'll make it up to you."

"It'll cost you," she smiled wryly.

I loved my girls. It was all set. Ami knew. The people knew. Almost time for business. A nearly forgotten sense of anticipation stirred my soul.

There are few things more exhilarating than leading men into battle. I recalled strapping on my kit…body armor, helmet, magazines, weapon, get it up and get it on, feeling invincible, powerful, united with men of steel from throughout the ages, a brotherhood of knights standing firm against the forces of darkness, preparing to descend from the blackness of night onto the unsuspecting heads of our enemies, visiting the wrath and judgement of a righteous and angry God unto the wickedness of evil men.

Seared in my memory is the gruff chorus, the guttural hum of engines starting, the synchronized whir of the rotors, the wind and driving sand, the airframe trembling in anticipation as the throttle was advanced to full open, the boys loaded, weapons and radios checked, and the uneasy silence on the net as the armored fist of martial supremacy waited to strike, poised to pounce and unleash hell and hellfire. I exalted. Rarely have I felt more alive, more real, than when marching lockstep with my brothers into a fight. Fear and uncertainty coupled with the quiet confidence of your own readiness, your own preparation, and the determined heart of the man on your left and right. Would I still have what it took?

Ever since I was a boy, I dreamed of fighting, leading men into battle and now it seemed, the Lord was offering me one more chance. I prayed that I still had what the men needed. I prayed for our victory. I prayed that I would honor the Lord, our God during this unsuspected, but welcome hour of testing. I'm sure that my heart might just beat right out of my chest.

1. Approaching Jericho

As Joshua surveyed the Wadi Qelt prior to the siege of Jericho, a man stood before him bearing a sword. Joshua approached the man and asked, "Are you for us, or for our adversaries?" The man responded, "No; but I am the commander of the Army of the LORD. Now I have come." At this, Joshua fell to his knees in the dirt and worshiped him saying, "What does my lord say to his servant?" (v. 14b) The man beheld Joshua and responded, "Take off your sandals... for the place where you are standing is holy."
Joshua 5:13-15b

First Things

Brave Rifle, Joshua stands as one of you, one of us, speaking directly from the pages of Scripture. The Bible, God's direct revelation to us, brims with the desperate sounds of battle: clashing swords, driving chariots, the battle cry of a thousand warriors. Blood is spilled. Nations rise. Kingdoms fall. Through it all, the Lord our God sits high and lifted up, searching hearts and testing men. (Romans 8:27)

Men perpetuate a damaging myth concerning the reality of Jesus Christ that resonates throughout the Church and western society. Popular accounts almost always depict Jesus in a certain way—soft and kindly, loving and forgiving, gentle and meek, likely pacifistic, definitely non-threatening. He would make no demands or do no harm.

In essence, modern day Jesus is a woman.

He exists, complete with European good looks and wind-blown hair, as a warm and fuzzy character, affable and sensitive, yearning in His heart for you. This faux Jesus heads an increasingly feminized Church, a church from which men flee in droves.

Men flee the contemporary love songs sung to Jesus. Men can scarcely stomach the endless line of pretty young men who croon incessantly about how they've fallen in love with Jesus. Men throw up in their mouths a little bit each time they sit before stylish and manicured preachers who prattle on about how Jesus just loves absolutely everybody.

We possess no need for a docile Jesus or a safe Jesus. The contemporary caricature of Jesus lacks the breadth of the true person of Christ.

Let us learn about war and those conducting it. Let us examine the corporate aspects of bloodshed and refuse enlightenment from the usual fathers, though their commentary may prove useful. Instead, let us seek illumination from the Warrior of all warriors, the Lion of the Tribe of Judah, the Lord, Christ Jesus.

He is a fierce warrior. He is a proud warrior. He is a powerful warrior, giving no quarter to His enemies, destroying nations, shattering kingdoms, ruling all with a rod of iron. He presently sits at the right hand of God on high until all of His enemies yet kneel as His footstool. (Hebrews 10:13, Psalms 110:2) On that day, He will return in power and glory and set all things right, settling accounts, reconciling the books forever. At His very name, at the name of Jesus, every knee will bow and every tongue confess that Jesus Christ is Lord, to the glory of God the Father Almighty. (Philippians 2:10,11)

Is this the Jesus you learned about in vacation Bible school?

Jericho and Ai

Consider Jericho. Joshua miraculously forded the Jordan River at the head of an army of 40,000. "On that day the LORD exalted Joshua in the sight of all Israel, and they stood in awe of him just as they had stood in awe of Moses, all the days of his life." (Joshua 4:14) Israel trembled with anticipation at the fulfillment of an ancient promise, a promise to the entire nation and indeed, to all nations.

Sitting just west of the Jordan River Valley, Jericho stood as the keystone to the entire conquest, the lynchpin that must fall should the rest of Canaan follow suit. Joshua, bold in his convictions and bearing the Ark of the Covenant at the head of his army, marched on Jericho while the king and his people braced for the onslaught.

Joshua wisely dispatched two spies to scout out the city who sought refuge in the home of Rahab the prostitute (Joshua 2)—here I'll refuse a generalized statement concerning the moral ambivalence of the average spy! Under interrogation, Rahab lied to the leaders of Jericho and sheltered the spies, allowing them to escape the city from a bedroom window as her home was built into the city wall. In exchange, the spies promised that Israel would spare her household from the coming slaughter.

The spies reported to Joshua, "Truly the LORD has given all the land into our hands. And also, all the inhabitants of the land melt away because of us." (Joshua 3:24) Canaan had observed Israel's advance for decades. For nearly 40 years, Israel maneuvered in the wilderness near Kadesh-Barnea, just south of Canaan. Frequent contact between them educated Canaan as to Yahweh's promise and Israel's intent to invade. Now, Canaan watched helplessly as Israel marshaled for war east of the Jordan River.

They heard of Israel's skirmishes with Arad and the Amorites. (Numbers 21) They trembled as Israel invaded Heshbon and defeated Sihon following his defiant refusal to grant Israel passage. (Deuteronomy 2) They received reports of the slaughter as Israel "devoted to destruction every city, men, women, and children." (Deuteronomy 2:34) Only the livestock survived. King Og of Bashan suffered a cruelly similar fate. (Deuteronomy 3) The Edomites deterred Israel forcing them to circumnavigate their territory, prompting the invasion of Canaan from the east. Jericho shuddered, "shut up inside and outside because of the people of Israel." (Joshua 6:1)

Joshua, at the behest of the commander of the Lord's army, prepared to tread the holy ground of conquest, intent on fulfilling God's plan and promise. Israel's warriors consecrated their hearts for the fight; Jericho awaited the sword.

This would not be a typical siege. At the LORD's command, Joshua assembled his army and selected seven priests, providing each with a trumpet to march before the ark. Forming a procession, they marched the perimeter of the city, first the men of war, followed by the priests and the ark, and then the rear guard. No one spoke as the priests continually blew their trumpets.

How strange this must have seemed to the people of Jericho. Expecting an attack, perhaps a blockade, maybe an assault on the ramparts, they instead watched this bizarre procession. Imagine the uncertainty. What were they doing? When would they attack?

Israel marched one time in procession around the city for six consecutive days. Jericho would not survive a seventh. On the seventh day, Israel circumnavigated the city seven times, at which point the army shouted as the priests blew their trumpets. Jericho's walls fell opening the city for invasion. Only Rahab and her household survived as "they (Israel) devoted all in the city to destruction, both men and women, young and old, oxen, sheep, and donkeys, with the edge of the sword." (Joshua 6:21) They killed the *livestock!* They killed the women and children, everyone except Rahab and her household, just as God ordered.

Militarily, Israel's rousing success at Jericho defied all expectations. Word of the victory soon spread and "so the LORD was with Joshua, and his fame was in all the land." (Joshua 6:27) Canaan trembled further. Emboldened, Israel looked to the city of Ai. Situated just west of Jericho, God once more delivered the city up to Israel just as He had Jericho. Following an initial setback, Israel slaughtered the entire population of 12,000. (Joshua 8)

As imminent as total victory must have seemed, Israeli momentum soon flagged. The Gibeonites avoided destruction, essentially deceiving Israel into signing a treaty allowing them to live and remain in the land. (Joshua 9) The remaining Canaanites fought tough, ceding ground but never yielding until the offensive finally stalled. Israel, for a number of reasons we will examine, failed to replicate the initial successes at Jericho and Ai.

Joshua's eventual death further derailed the invasion and the conquest concluded incomplete, giving way to the historical time of the Judges. Israel's hopes of peace waned alongside their wavering resolve. The failed invasion would haunt Israel for the remainder of its existence. Perpetual warfare plagued the nation and eventually a series of Gentile kingdoms would subjugate Israel until its destruction in A.D. 70, all stemming from the unfinished conquest of Canaan.

Main Effort

> *Now these are the nations that the LORD left, to test Israel by them, that is, all in Israel who had not experienced all the wars in Canaan. It was only in order that the generations of the people of Israel might know war, to teach war to those who had not known it before. These are the nations: the five lords of the Philistines and all the Canaanites and the Sidonians and the*

*Hivites who lived on Mount Lebanon, from Mount Baal-hermon as far as Lebo-hamath. They were for the testing of Israel, to know whether Israel would obey the commandments of the LORD, which he Commanded their fathers by the hand of Moses. So the people of Israel lived among the Canaanites, the Hittites, the Amorites, the Perizzites, the Hivites, and the Jebusites. And their daughters they took to themselves for wives, and their own daughters they gave to their sons, and they served their gods.
Judges 3:1-6*

War serves many purported masters. From antiquity, war served the political and economic ambitions of the nation-state and the subsequent defense against rival nation-state aggression serving their own political or economic ambitions. War frequently served religious purposes.

Religion and warfare share a distinguished and sordid historical cohabitation. Men have shed blood on behalf of God for as long as any can remember. Yet, secular examinations of the business of war have neglected examining war with the assumption that a particular religion might be true. They've examined war from an external vantage, treating religion as a competing ideology that might drive the impassioned zealotry of certain men. In doing so they have overlooked examination from the actual viewpoint of said ideology.

We will address and examine the issue of war from a decidedly spiritual perspective. The scrupulous student of the Bible proclaims all things as spiritual. All things possess a spiritual component. All people possess an immaterial aspect. Spirituality governs all things, all relationships, all interactions, and all issues. This includes war. Would not a spiritual examination prove useful, essential even?

I don't intend, in this work, to make a defense of the Christian religion. Though evangelism is not my primary objective, nevertheless as a believer, at some level, evangelism governs my intent. Why would anyone adhere to a system without proclaiming it to be true? This work contains, intertwined with applicable battlefield accounts, certain relevant aspects of my own journey from unbelief to belief. For a more exhaustive exploration of the truth claims of Christianity, explore Lewis' *Mere Christianity* or McDowell's *More than a Carpenter*.

Perhaps you adhere to another faith. Perhaps you have no religious faith, at least any that maintains bearing upon the reality of your life. I still contend that the enclosed examination might prove useful to you. I could tell you that God's common grace stipulates that, "he makes the sun rise on the evil and on the good, and sends rain on the just and the unjust," (Matthew 5:46b) that His grace benefits all people, not just Christians. However, the discerning skeptic would instantly recognize the circular logic contained herein.

We will search the Scriptures for what He says concerning war and I'll leave you to draw whatever conclusions you may. Perhaps, following examination, you may decide to discard these conclusions. Perhaps you will come to recognize that all of creation speaks to the existence of a Creator, even man at his most hostile…especially man at his most hostile.

We will seek a deeper motive concerning war, a truer conviction than previously asserted. We'll seek to define a distinction between God's kingdom and worldly kingdoms and associated methods of establishing these disparate realms. We'll meditate on the idea that all things serve the purposes of God, including war. We'll examine God's intentions for the business of war as we glean from our text the following observations:

1. God tests through war and,
2. God teaches through war.

I pray that the Holy Spirit would teach us, enlighten us, and reveal to us truths of God's word contained in the Bible. God bless you in your pursuit.

Chapter 2

2. Context of the Examination

The rigor of the Christian existence stuns the newly-minted believer. Upon my conversion in 2005, I never imagined the sheer level of inclusiveness insisted upon. I had intended, now that salvation had been "taken care of", to return to my previous existence. Almost immediately, the spirit and the flesh began warring with one another. They remain locked at the horns to this day.

Totality of Surrender

The Bible calls the believer to "rejoice always, pray without ceasing, give thanks in all circumstances." God calls the believer to take up the cross daily, to crucify the flesh daily. Jesus demands absolute surrender of one's life unto His authority. (1 Thessalonians 5:16-18, Galatians 5:24, Matthew 16:24) Peter, speaking on behalf of the Apostles, declares, "See, we have left everything and followed you." (Mark 10:28) The Apostles abandoned everything to follow Jesus—their vocation, their families, their security.

Jesus even calls the believer to abandon that which he loves. To the man desiring to first bury his father, an exceedingly important thing, Jesus declares "Leave the dead to bury their own dead. But as for you, go and proclaim the kingdom of God." (Luke 9:60) To the man desiring to hesitate for a moment and bid his family farewell, Jesus responds, "No one who puts his hand to the plow and looks back is fit for the kingdom of God." (Luke 9:62) These declarations, when taken at face value, clash with the selfish, innermost recesses of the residual self. (Romans 6:6)

Jesus' words in Luke chapter 14 force the issue.

> *If anyone comes to me and does not hate his own father and*
> *mother and wife and children and brothers and sisters, yes, and*
> *even his own life, he cannot be my disciple.*
> *Luke 14:26*

Jesus calls the believer to *hate* those he should otherwise love, a difficult prospect demanding an explanation. Jesus does not actually intend hatred toward those He elsewhere calls us to love and to serve. This passage and those previously mentioned call the believer to fully surrender the totality of one's existence to the authority of Christ.

Jesus does not actually call the believer to hate his family. Speaking in hyperbole, He posits that the believer's love for Him must be so extreme that in comparison, his feelings towards his family would be as hatred. These ideas did not resonate well within my newly formed heart as a young believer.

I recall a particular encounter early in my walk as God wrestled more and more of my life into His purview. As all that belonged to me continued to dwindle, I shouted in frustration at Ami, "It can't all be about God can it?!" God, it would seem, intended exactly that.

From the Soul

Paul writes, "Whatever you do, work heartily, as for the Lord and not for men." (Col. 3:23) The verse literally reads that whatever you do, do it "from the soul." Do nothing half-heartedly. Do everything as if you serve the Lord Jesus Christ. If you are a plumber, clear clogged toilet bowls as if working for Jesus. If you are a dentist, drill cavities as if working for Jesus. If you are a barista, brew espresso as if laboring for Christ Himself. Whatever you do, do it from the soul, with all your heart, unto the Lord.

Could this apply to warfare as well? Could one wage war from the soul? Could one wage war unto the Lord?

The Bible calls us to be consumed by the Lord. Matthew records Jesus answering questions concerning the greatest commandment. He responds, "You shall love the Lord your God with all your heart and with all your soul and with all your mind." (Matthew 22:37) Here Jesus referenced the great *Shema'*, the uniquely monotheistic Hebrew statement of faith, "Hear, O Israel: The LORD our God, the LORD is one. You shall love the LORD your God with all your heart and with all your soul and with all your might." (Deuteronomy 6:4,5) Jesus' reference speaks to a totality of the believer's love for God.

Love God with all that we have, all of our existence. Surrender all things to Him, to His will. These directives confront those wishing to confine their Christianity to a certain sphere of their existence. I assert that my own struggles in this regard likely resemble those of countless brothers. For the believer, the remnants of the dying flesh war with the Spirit concerning dominion until glorification, the final settling of the accounts in death. The unbeliever wrestles yet a different animal, wages a different war.

Interpretations

As we flesh out the idea that God tests and teaches through war, a thorough understanding of *means* becomes necessary are we to proceed. How do we draw our conclusions? On what basis? Absent a solid grasp of methodology, Judges 3:1-6 will ring hollow.

God has revealed Himself to us in the pages of Scripture. How we interpret Scripture drives our conclusions as indicated by the fact that many arrive at different conclusions about God from the same text.

Context proves decisive in any endeavor. I am a biblical literalist in my interpretive philosophy. I take what the Bible says literally unless it provides a reason to do otherwise. An example of this would be in Galatians when Paul speaks of the covenants,

> *Now this may be interpreted allegorically: these women are two covenants.*
> *Galatians 4:24*

Paul references the true account of Sarah and Hagar but provides in the following text an allegorical interpretation concerning the covenants.

The rich literature of the Bible brims with all manner of literary styles: historical narrative, genealogy, statutes/laws, parable, prophecy, poetry, allegory and apocalyptic writing. When we read the Bible, we must accurately assess the style of literature as we seek an accurate comprehension.

When Jesus presents a parable, we understand this to be a fable-type story generated for the purposes of teaching a lesson. As Jesus described the dishonest manager, we don't necessarily conclude that Jesus knew a dishonest manager. Perhaps He did, but we concern ourselves with the lesson from the parable. Sometimes Jesus alerts us to the fact that He speaks in parable. Most of the time He does not.

The book of Revelation is the most difficult book in the Bible to understand; after years of study I still tremble at the thought of preaching through it. Everyone desires to know what happens at the end, so what do they do? They flip to Revelation and start reading. Instantly Scripture confronts them with surprising images, confusing scenes, and haunting characters. Will angels literally sound trumpets, pour out bowls, and break seals during a future time of judgment? Perhaps, but I know that Revelation, as an apocalyptic writing, utilizes extreme forms of vivid imagery and intricate symbolism to make various points. Curiously, you'll find those who waffle on other literal aspects of the Bible resolutely declaring future tribulations and anti-Christs, beasts and dragons and the like, all from a cursory reading of Revelation.

As with any subject, context is decisive. What is the literary context? People turn to Revelation to understand end times events, but skip the vast wealth of previous writings concerning end times—Ezekiel, Daniel, and even Jesus who offers much on the issue in texts such as Matthew chapters 24 and 25.

Historical context provides an equally important factor. What is the historical context of the text? When did it happen? Who wrote it? To whom is the author writing? What is the occasion, what is prompting the author to write? What does the author intend? What was happening at the time?

Neglect of literary and historical context frequently leads to misinterpretation. Consider our tendency to elevate ourselves, to consider ourselves as the focus. Poor biblical interpretation allows us to do exactly this with Scripture, apply texts to ourselves that the author never intended.

Jeremiah 29:11 provides an outstanding example of this,

For I know the plans I have for you, declares the LORD, plans for welfare and not for evil, to give you a future and a hope.
Jeremiah 29:11

Many take this verse and declare it a promise for themselves or maybe a friend who struggles in a particular way. "Look, God has great plans for you! Just look at Jeremiah 29:11. God wants to prosper you, just look, it says it right here in Jeremiah. He has a great plan for you!"

With the noblest of intentions, they miss the mark concerning the application of this text. What about the martyrs, the countless number who were burned at the stake or drowned for their faith? Did God not have a wonderful plan for their life? What about the Army private who gave His life to Christ and was killed on a remote mountain in Afghanistan less than a week later? Did God not have a wonderful plan for his life? What about the Apostles? They were tortured, crucified, beaten, and burned, all but John dying a martyr's death. Did God not have a wonderful plan for their life?

With assurance, I agree that God definitely has a plan for your life. With even more assurance I'll declare that this plan likely includes suffering of some kind, not what most would call wonderful and prosperous. Jeremiah wrote to the exiles in Babylon, urging them to retain hope. He exhorted them to continue to have children and encourage their children to have children, that He will one day return them to the Holy Land and restore them. Jeremiah declared a blessing upon the children of the exiles, since those he addressed would likely die prior to the return. You want to believe that God has a wonderful plan for your life, plans to prosper you and He may. This text just doesn't support that idea.

People apply Matthew 18:20, "For where two or three are gathered in my name, there I am among them" to mean that somehow, when people gather to pray, God draws near as if He magically hears these prayers more clearly. Scripture speaks to the omnipresence of God. He already exists everywhere, in all places. Corporate prayer pleases God, but this text does not support the notion that a gathering produces a special presence or manifestation.

In this passage, Jesus concludes a diatribe on church discipline, confronting the sin of a brother in Christ, gently seeking restoration. After an initial one-on-one confrontation yielding no repentance, the offended party brings in witnesses and eventually the church. Church discipline honors God and requires the presence of two or more and God assures the believer of His presence during these encounters.

I saw Philippians 4:13 on a workout shirt the other day, "I can do all things through him (Christ) who strengthens me," as if the presence of Christ would help knock out another rep or shave a few seconds off a time. John 14:13, "Whatever you ask in my name, this I will do," spawned an entire cottage industry of prosperity preaching. "Proclaim it, speak it into existence, ask it in the name of Jesus and it will be done!"

Unfortunately, these misapplications enslave a multitude and deceive many into false understandings about God. When reality fails to meet their faulty biblical understanding, they turn to God in anger or even worse, abandon Him entirely.

Context is key, decisive. To fully understand any Scripture we must ascertain its literary and historical context and interpret it through the lens of the entirety of God's word. Thus the historical-grammatical hermeneutic most faithfully interprets the Bible. We first seek the original meaning, translate it to our context, and only then seek application. The plainest reading and most simple understanding of any passage is always best. As Jesus said to Nicodemus,

If I have told you earthly things and you do not believe, how can you believe if I tell you heavenly things?
John 15:12

I think of Occam's razor, a principle that declares that all things being the same, the simplest answer is usually the correct one. The more vigorously one must develop a belief or a doctrine, the less likely it is to be true.

The historical-grammatical method requires that I err on the side of the word of God. I may not comprehend exactly how something in the Bible may be true, but if it's in the Bible I declare it to be true. It just so happens that all of my studies have done nothing but reinforce my belief in the Bible and this method of interpretation, especially considering that I am a math/science type of person. I was raised in a family of engineers. Methodical query is in my genes.

Heresy beckons. As we stray from the historical-grammatical hermeneutic, all things lose their footing; foundations crumble. Modern attempts to square sins such as homosexuality with the Bible start with a dissolution or a misuse of this method. Legalism, unbiblical church polity, unscriptural practices such as ordaining female pastors—they all start with a departure from proper biblical interpretation.

We will cling to the literality of the Bible and contend with whatever outcomes that may yield, particularly those outcomes that pertain to war.

Unanswered Questions

I've always been a student of war. As a teenager, while my friends decorated their bedroom walls with images of scantily clad pin-up girls or popular rock bands (as this was the 80's, they both sported the same hairdo), I displayed a single poster bearing images of Patton, Eisenhower, Lee, and Grant. It read, "At West Point, much of the history we teach, was made by people we taught." The timeless bonds of historical military leaders captivated my imagination from my earliest days.

I consumed war movies—*Platoon, Hamburger Hill,* and *Full Metal Jacket* being a few of my favorites and I impugn anyone who liked Sergeant Elias over Staff Sergeant Barnes. Eventually, I graduated from pretending with sticks, toy guns, and fireworks to actually preparing to walk in my heroes' footsteps.

As much as I've always been a student of war, I always felt called to be a soldier. I never once required encouragement from my parents to attend the military academy or join the armed forces. To me, nothing seemed more honorable or more worthy, and I longed for the day when my mettle would be put to the test on the battlefield. My uncle epitomized cool to me. A West Point grad who flew Hueys and Mohawks during separate tours to Vietnam, he was what I strove to emulate.

Decades later, I accomplished what I originally intended, never once wavering. It was as if my days were already numbered, ordained beforehand, when a curious and completely unanticipated thing happened. I joined another battle, the actual battle.

In March 2005, I raised my hand and surrendered my life to Jesus Christ. Lest you misunderstand, I never once pursued Him. I was brought before Him. Circumstances—sovereignty actually—delivered me unto Him, and for the first time that I recall I heard the actual good news of the Gospel of Jesus Christ in all its power, all its authority, and with all of its teeth. God confronted me with my sin and made me aware of my rebellion against Him. The Holy Spirit convicted my heart of this sin. Though I did not seek Him, He sought me and I possessed no choice but to fall to my knees in surrender.

I repented of my sin and believed on the Lord Jesus and was made a new creation in Christ. At the time, I possessed scant understanding of what that actually meant, of the magnitude of this new reality that I now possessed, this new prism through which I must view the world. I scarcely understood that *all* things had now changed *irrevocably*.

Yet, I had a problem. I was a soldier in the Army preparing to deploy once more to the killing fields of Iraq to close with and destroy the enemy. What should I do with this new and unexpected aspect of my existence, my faith?

My new Master, Jesus, said some fairly profound and revolutionary things. From the Sermon on the Mount,

Love your enemies and pray for those who persecute you,
Matthew 5:44

If anyone slaps you on the right cheek, turn to him the other also,
Matthew 5:39b

Love your enemies, do good to those who hate you,
Luke 6:27

Be merciful, even as your Father is merciful.
Luke 6:36

What to do with these things? How could I reconcile these newly discovered revolutionary teachings of Jesus with what my nation now called me to do?

The movie *Born on the Fourth of July* chronicles the true story of Ron Kovic, a motivated and idealistic young man who enlists in the Marine Corp to fight the communists in Vietnam. On the battlefield, the stark reality of the conflict brutally confronts his idealistic and naive impressions. He witnesses war crimes and the slaughter of children. He accidentally engages and kills a comrade during the fog of battle and is himself grievously wounded, paralyzed from the chest down. Much of the film documents his post-war struggles with guilt, shame, and purpose, exacerbated by the demons of alcohol abuse.

In one particular scene, Kovic returns drunk from a night out, ranting like a lunatic about his parent's home, crying and raging about the atrocities he witnessed and even took part in, lamenting his lost manhood, while angrily brandishing the crucifix of Christ against a nonexistent God.

"They told us to go," Ron agonizes.

"Yes, that's what they told us," his mother cries in return.

"Thou shall not kill, mom. Thou shall not kill women and children," as the sheer guilt and agony of what he had seen and done tore at his soul. "Thou shall not kill, remember? Isn't that what you taught us?"

I knew this to be an idealized account for the sake of box office revenue but was this to be my fate? Was I doomed to be torn apart by the inevitable clash between my duty as a soldier and my new worldview, my new convictions? All manner of inquiry tore at my mind.

> Should a Christian fight in war?
> What was my duty?
> What of Jesus' teachings? Can they be reconciled?
> What does God think about war?
> What does the Bible say?
> What do we do with the Biblical accounts of the slaughter of women and children?
> What about atrocity? War crimes?

Just over a decade later, I have firmly reconciled these convictions, seared them into my conscious. I pored over the pages of Scripture, consulted with pastors, listened to teachings, and read numerous books. The journey of my faith led me to explore numerous and much more provocative issues than war: issues concerning grace and redemption, issues concerning conduct, issues concerning future things. Ever the student, I must now return to war.

I assure you of the inevitability of these queries. All men run a collision course with the Almighty. You may evade a bit longer but at some point, your pursuits, in this case your engagement in the business of war, will drive you to a confrontation with the reality of God. I ask that you might reconcile in advance, while you still possess the luxury of time. Perhaps He already drives you to your knees.

Would you know the truth?

Chapter 3

3. First Things, Hardest Things

We'll proceed to the bloodshed soon, but please understand that our examination falls short if we do not first establish a functional grasp of the relationship between God and things, in general. Even as we seek the relationship between God and the thing that is war, we remain mired until we grasp how God relates to all things. Only then may we drill further into the notion of how God relates to war.

A broad understanding of the scope and magnitude of the person of God proves initially decisive, effectively setting the conditions for deeper queries.

The sheer magnitude of God prompted Paul to exclaim,

Oh, the depths and riches and wisdom and knowledge of God!
How unsearchable are His judgments and how inscrutable his
ways!
Romans 11:33

This followed an exhaustive dive into perhaps the deepest biblical truths in Romans chapters nine through 11, truths concerning grace, predestination, and election. The magnitude of God prompted the prophet Isaiah to declare,

For my thoughts are not your thoughts,
neither are your ways my ways,
Declares the LORD.
For as the heavens are higher than the earth,
So are my ways higher than your ways
and my thoughts than your thoughts.
Isaiah 55:8-9

The LORD's ways transcend man's ways. They differ wholly, completely. God says to the psalmist,

... you thought that I was one like yourself. But now I rebuke you and lay the charge before you.
Psalm 50:21

God rebukes the psalmist that he might consider himself similar in any way. The magnitude of God befuddles the finite minds of men. You may ask then, "What use could there be in studying God?" Much! The pages of Scripture, which are God's revelation concerning Himself, yield more than the finite mind could consume in a lifetime. So rich and glorious are the depths of God.

The Incomprehensible

God is unsearchable, defying comprehension. He is vast and immanent, transcendent yet personable. He is uncaused and uncreated. His very being defines existence.

When confronted at the burning bush, Moses asks of God, "If I come to the people of Israel and say to them, 'The God of your fathers has sent me to you,' and they ask me, 'What is his name?' what shall I say to them?" (Ex. 3:13) God responds, "Say this to the people of Israel, 'I am has sent me to you.'" (v. 14b) His Hebrew transliterated name יְהֹוָה is YHVH or Yahweh, literally rendered "I am".

He is omniscient, omnipresent, and omnipotent. Were we given a thousand lifetimes to seek Him, to immerse ourselves in God's special revelation to man, we might just begin to comprehend what it is we do not yet understand about Him. Indeed, the deeper I dive into the knowledge of the Holy, the more aware I become of just how shallow my previous excursions have been.

If we are to understand the relationship between God and war, then we must first attain some understanding of God. How else could we go about it? You may be tempted to forgo the loftier theological ambitions of this work and head immediately to the tangible. Skip the God talk and turn directly to the bloodshed. Allow me to caution you against such a notion.

Perhaps you've fought, perhaps you've seen the elephant. Perhaps you are a student of war such as I and you've reached some previous conclusions. Allow me to assert that unless your conclusions somehow account for the intangible, the spiritual, they lack validity.

The fear of the LORD is the beginning of knowledge.
Proverbs 1:7a

We can never fully comprehend the thing that is war apart from the fear of the LORD, and it is knowledge of Him that generates godly fear.

Maybe you serve a different god or maybe you possess no faith. Perhaps you pursue this work in a secular regard seeking only another vantage point with which to bolster, refute, or confirm an already well-entrenched stance.

Historically, numerous students and practitioners of war claimed no affiliation with the God of Christianity. One could argue that many powerful practitioners of war forsook matters of faith entirely. Consider the godless regimes that generated the brutal efficiency of the Wehrmacht or the harsh but effective pragmatism of the Red Army. Especially when considering the godless, a study of the God of the Bible proves fruitful and even critical.

Initial Incursion

We will first pursue a pedestrian comprehension of an enigmatic issue, the providence of God. This issue highlights the complex, delicate, and frequently misunderstood relationship between God and His creation. Understanding God is one matter, understanding His creation yet another, reconciling the relationship between them, a still more ambitious endeavor.

A modern helicopter showcases beautiful complexity and stunning intricacy. At a hover, tens of thousands of pieces, many moving at insanely high velocities, cooperate and interact with perfect harmony. External factors such as jet fuel, pilot input, and the physical environment collude with the machine. Appropriately addressed, these culminate in the harnessing of certain aerodynamic principles that yield a thrust vector running exactly opposite in direction and magnitude to the pull of the gravitational force upon the center of gravity of the airframe. The aircraft hovers.

I may understand, with reasonable certainty, how the engine works. I can ascertain, without too much difficulty, the function of the hydraulic system. I can be trained to push the right buttons, to move the flight controls in the appropriate manner. However, the fusion of all of these into a unified operation requires synergistic thought beyond that which is ordinary. Interaction between these tens of thousands of parts, humans, fuel, and the environment demands higher order cognition. It is such with God and His interaction with His creation.

Though I will never fully grasp God, my language might indicate a level of understanding. As an example, I may speak of his immutability and say with certainty that yes, God never changes. He cannot change. He is the same today as He was yesterday and will be tomorrow, as immutable as He could possibly be. The Bible declares it, I believe it. Examine Malachi 3:6, "For I the LORD do not change," or Numbers 23:19, "God is not man...that he should change his mind," or Psalm 102:27, "but you are the same, and your years have no end."

Yet in places the Bible suggests that God does exactly that, change. From Genesis, the pervading wickedness of man suffers the LORD and, "the LORD regretted that he had made man on the earth." (Genesis 6:6a) He regretted making man, regret implying a change in attitude. Centuries later, Abraham convinced God not to destroy Sodom. He bargained with God, seemingly able to change His mind. (Genesis 18) During Moses' absence at Sinai, Aaron and the people produced a golden calf to worship, clearly a violation of the foundations of the law. God declares in anger, "let me alone, that my wrath may burn hot against them and I may consume them." (Ex. 32:10a) God intends to destroy them. Moses intercedes and "the LORD relented from the disaster that he had spoken of bringing on his own people." (Ex. 32:14) Here again, God appears to change.

The concept of immutability *seems* easy and straightforward until overlaid upon His creation and the events of life.

Thus, we must reconcile this relationship, the interaction between God and creation labeled *providence*. We must delineate governing principles. We must define the terms. As we seek command of the theology of war, a less than marginal grasp of the relationship between God and creation will detract from this endeavor.

We'll confine our discussion to three important topics:

  1. *Sovereignty,*
  2. *Providence, and*
  3. *Concurrence.*

Each of these could command a voluminous exposition. We will make our necessary assertions and move forward. However, we will quickly arrive at, but not dwell upon, a fundamental unknowable aspect of God.

Many qualities of our infinite God prove difficult to reconcile yet we can retreat into the fact that they just are. For instance, God is infinite, unbound by the constraints of time. We can *say* this, but as we are yet bound, we cannot *know* what this actually means. We still declare it.

However, in resolving the interaction between God and His creation, one inevitably arrives at a position of not knowing, of not even being able to define in familiar terms. We will quickly find ourselves declaring seemingly contradictory truths and if we are not careful, affirming dissonant notions, as if the Lord has no regard for logic.

Theologians from all days concur that the things we will soon examine just are. Reality requires that they be true. Thus, we may join Nicodemus in lamenting, "How can these things be?" (John 3:9) as the sheer incomprehensibility of these truths overwhelms our finite and clouded minds.

War is, amongst all of the other ways that men have sought to define it, an interaction between God's creation, both persons and things governed in a yet-to-be-described fashion by God. Thus, this rudimentary understanding of sovereignty, providence, and concurrence will prove decisive in yielding mastery of the object of our pursuit, the theology of war.

Preparing the Battlefield

We live in the 21st century, an interesting time when it is permissible to believe whatever you want as long as you don't declare it to be true, particularly for anyone other than yourself. Believe what you want but please do not declare it to be true for others. Contemporary western society adores subjectivity, absolute truth being a decidedly obsolete and priggish notion.

Against this backdrop of subjectivity, we'll first declare some objective truths. These presuppositions alone have generated conflict from the beginning, consuming kings and nations. Such is the magnitude of God that even an exposition of necessary presuppositions proves a colossal undertaking.

Though this work could never do them justice, I strongly urge you to examine them, tear them apart, and immerse yourself in them. I urge you to reject my word and see for yourself. As it is, we must declare them and move on or we will never reach our intended objective.

1. **The Necessity of this Endeavor.**

You may inquire as to the necessity of a theology of war. Why must we even make this attempt? This query possesses much merit as countless greater minds have engaged the subject of warfare and its definition in a comprehensive fashion. Why must we examine warfare from a decidedly religious perspective? A Christian perspective? Is it even useful?

Every man possesses and applies ideas about God to whatever they do, whatever activity they participate in. Perhaps you are an atheist and indignantly state, "Well for me, certainly not!" Doesn't this presupposition about God, His inexistence, undergird your thoughts and actions?

Those who most vehemently deny His existence often serve as the most vocal catalysts for discussions concerning the same. He overshadows everyone, all things, such that His presence affects even the thoughts of those who would otherwise deny Him. This speaks not only to His very existence but the necessity of discussion and definition concerning His existence.

The Bible claims that everyone knows that God exists deep in their hearts. Speaking to the unrighteousness of men, Paul writes,

> *For what can be known about God is plain to them (*the unrighteous)*, because God has shown it to them. For his invisible attributes, namely his eternal power and divine nature, have been clearly perceived, ever since the creation of the world, in the things that have been made. So they are without excuse.*
> *Romans 1:19-20*

Did you catch that? Creation plainly speaks to God's existence. Everyone knows that God exists, as revealed by His creation. All men know this, yet many do not acknowledge it. Many rage against this notion and why, because to acknowledge God is to acknowledge authority and ultimately, accountability.

Contrary to many flawed objections to Christianity, men are not condemned for rejecting Jesus. They are condemned for rejecting God. Paul further states that they "exchanged the truth about God for a lie and worshiped and served the creature rather than the Creator." (v. 25) Could any criticism more accurately define the observed wickedness in the hearts of men? Men reject God. They worship the creature rather than the Creator.

Men know that God exists yet do not desire the burden of His existence. This denial governs thoughts and actions, defines practices. We want to worship the created, not the Creator.

Even those who do acknowledge the existence of a higher power frequently generate a god of their own understanding. Men create never-ending sources of erroneous theology, often derived from conjured and fictitious gods of their own creation. In western society, men generally worship *The Great Whatever* or *The Big Man Upstairs* while women and the young seem to view God as a grandfatherly-type character who just wants young people to have a good time.

I've always been an avid reader, and as a young man I ravenously consumed Stephen King novels. I actually developed much of my agnostic theology from his novel *The Stand*. As silly as such a notion may be, for decades an academic, godless mindset governed and darkened my existence, driving my actions. My concerns originate from my personal experience.

Our understanding of God always drives our actions. Orthodoxy, a correct understanding of God, always precipitates orthopraxy, right practice. My darkened views of God and His existence generated darkened living and action on my behalf. I did good things and maintained a surface level morality. Yet, a rottenness of soul, an internal decay, tainted every decision I made, every thought I considered, and every action I took.

We must rightly define God and our beliefs, correctly keeping the end in mind. We seek right practice and with this regard, that we might understand and rightly engage in war. If we are to properly understand war and our relationship to it, we must understand God correctly. If we are to practice war correctly, we must master the theology of war.

2. The Bible is the Inspired, Inerrant Word of God.

I find it telling that man has developed an entire school of thought, that of biblical criticism, to slander the historicity and reliability of the Bible. Pseudo-scholars have sought to deconstruct the solid foundation that is the Bible for centuries and with increasing voracity, brick by brick. Amazingly, the more they dig, the deeper they discover that the foundation runs.

Every Christian church maintains a statement of faith, at least every one that I've seen. A church's views concerning the Bible provide great insight into their orthodoxy. Personally, I would never attend or promote a church that didn't have, at or near the top of the list in its statement of faith, something concerning the inerrancy of the Bible.

My church's statement opens with,

> We believe that the Bible is the inerrant, inspired, Word of God.
> It is the standard for all that we believe and practice.
> Furthermore, we affirm the inerrancy, authority, sufficiency, and clarity of Scripture.

The Bible is the most amazing book that mankind has ever produced and it's not even close. The writers, over 40 of them, spanning 2000 years, under the inspiration of the Holy Spirit, produced a perfectly congruent revelation of the Lord Jesus Christ that chronicles the salvation history of man. Consider that fact alone. Writers from different eras, different epochs, and different cultures composed this perfectly consistent work with no editor, no publisher, and no oversight committee directing the process.

It is a closed canon, not open to addition or subtraction. Contrary to some misguided academia, the early church universally and independently assembled the Bible into the collection of books we now possess, largely in response to early Gnostic heresies. No council or organization directed the codification of the Bible but confirmed what early believers had already decided for themselves again, with no governing oversight.

The manuscript evidence for the Bible dominates any other ancient work. Examine for yourself the thousands of manuscripts, many of which date to within a century or even a generation of the actual events chronicled. No other ancient work is even in the same hemisphere when it comes to the reliability of the Bible as far as what the original authors intended. The Lord even provided a contemporary affirmation of the ancient Old Testament as the Dead Sea Scrolls, discovered between 1946 and 1956, verified the absolute reliability of the documents already on hand.

I recently spent a few minutes with a young man who had some questions concerning the Bible, questions you may have. How do we know the Bible is true? How do we know that the documents we currently possess are what the writers originally intended? After less than 20 minutes of discussion he left, remarkably convinced. You see, a certain multitude doesn't want the Bible to be true because of the natural implications that such truth yields. No one, and I do mean no one, seeks to undermine any other historical work with nearly the same veracity, even other religious works. They reserve such attempts for the Bible.

In assuming the inspiration and inerrancy of the Bible as one hundred percent true and free from error, we must now establish a single, theological presupposition that nevertheless provokes all manner of human reaction, frequently negative, often violent.

3. The Deity of Jesus.

In the 4th century A.D. an Alexandrian presbyter named Arius who had studied under Lucien of Antioch generated a firestorm by declaring, with respect to Christ that, "there was when he was not." He denied the orthodox view of Jesus as the eternal Son of God and deemed Jesus to be subordinate to God, less than God, a demigod of sorts. Arius denied the triune God of orthodox Christianity. The proliferation of his teachings prompted

Constantine to call the very first ecumenical council, the Council of Nicea, which affirmed the orthodox view of Jesus and condemned Arius as a heretic.[1]

The Arian heresy did not die with Constantine's attempts to eradicate it. The denial of the deity of Christ is a heresy with surprising longevity. The Jehovah's Witness cult draws much of their theology from Arianism in the denial of the deity of Christ. They share a mutual Christology, primarily that Jesus is a created being, not of the same substance as God. Unitarians likewise share heretical Arian theology as do other cults. Scripture affirms that this denial will persist until the return of Christ and the subsequent abolishment of sin.

To develop an accurate view of God and how He relates to His creation, we must cling to an accurate Christology. In understanding the nature and character of God we will frequently reference New Testament descriptions of Jesus, as He is not just the central aspect of Christianity, but God incarnate.

Jesus is the God-man, fully God, fully man. The *hypostatic union* defies our intellects but is a natural progression of an orthodox understanding of the Trinity. Is Jesus God? Yes. Is Jesus man? Yes. He possesses all of the attributes of the Father though He shelved the independent use of some of these during His incarnation in a manner that is not entirely understood. (Philippians 2:6,7)

Paul's letter to the Colossians defines Him as "the image of the invisible God," (Col. 1:15) while the author of Hebrews declares Jesus as "the radiance of the glory of God and the exact imprint of his nature." (Hebrews 1:3a) God has revealed Himself to us in the person of Jesus Christ. As men touched Jesus, heard Jesus, lived with Jesus, they provide to us, through the Bible, a revelation of God. We must consider the reality of Jesus before we might ever attain a working knowledge of God.

4. The Necessity of this Endeavor (Revisited).

I'll maintain a certain pragmatism that I seek to expound upon in the latter parts of this work. That a healthy theology will doubtless yield a right application, a right conduct of war remains consistent with traditional Scriptural teaching concerning practical application of sound theology. That

---

[1] V.L. Walter, "Arius, Arianism," in *Evangelical Dictionary of Theology*, 2nd Edition, ed. Walter A. Elwell (Grand Rapids: Baker Academic, 2001), 95-96.

is the natural conclusion: right theology leads to right conduct with regard to any practice, including the practice of war. What of effectiveness aside from rightness?

As students of warfare, we seek the same end, effectiveness and efficiency at the conduct of war. Did Sun Tzu write for his own personal edification? Did Clausewitz and Jomini pontificate for financial gain? No, they all sought understanding for the ultimate purposes of national victory, that their teaching and insight might, rightly applied, yield victory.

It is with the deepest convictions, after years of study and yet more years of application, that I resolutely and confidently declare that orthodox biblical application will necessarily yield the most effective army possible, an army that achieves the most success on the battlefield. Later, we'll examine the alternative, the inherent dangers of a godless army. I pray that godliness and orthodoxy might re-permeate the fabric of this nation's army, for which I've labored all these years. A godly army will more likely defeat the evil and godless forces of this world that would otherwise destroy our cherished way of life. Pragmatism drives this pursuit though a revival within the ranks would serve far greater purposes.

Though I've clearly done inadequate justice to these presuppositions, I'll declare them firmly established, an opening salvo if you will. Again, I urge you with all gravity to pursue them to their appropriate end. Now, though, the study at hand beckons.

In the manner of Lee at Gettysburg, we'll initiate a frontal assault only instead of the Union lines, we'll lay siege to the mind of God. I pray our preparatory fires prove more decisive than those afforded Pickett during his ill-fated charge.

Sovereignty

We'll first declare the sovereignty of God as we simultaneously deny the feckless, weak, and marginal god crafted in the minds of fickle men. We will define sovereignty as,

> *The biblical teaching that God is king, supreme ruler, and lawgiver of the entire universe.*[2]

---

[2] F.H. Klooster, "Sovereignty of God," in *Evangelical Dictionary of Theology*, 2nd Edition, ed. Walter A. Ewell (Grand Rapids: Baker Academic, 2001), 1131.

The infinite second and third order implications from this foundational characteristic has occupied the minds of theologians since the Apostle Paul. Church fathers such as Origen and Augustine, Reformers such as Luther and Calvin, and even modern theologians such as Piper and Sproul all sought to reconcile this truth and all that it demands. What exactly does it mean and what are the implications?

Know that God is sovereign ruler over all things, not some things, all things. He is the very source of authority. He defines authority as all authority derives from Him. Nothing retains authority notwithstanding His purview. All things, people, and actions in time past, present, and future succumb to His rule. Nations bow to His sovereignty. Beasts of the field and birds of the air kneel to His supremacy. The weather, the oceans, the stars, the furthest imaginable galaxy to the delicate and impossible intricacies of DNA all submit to His authority.

Nothing happens apart from His ordination, not a single thing. No one speaks a word apart from His authority. No one fires a shot apart from His authority. No one dies and no one lives independent of His will and His authority.

He is the Creator of all things and the Sustainer of all things. Examine the pages of Scripture and come to terms with His absolute authority.

You rule over all the kingdoms of the nations. In your hand are
power and might, so that none is able to withstand you.
2 Chronicles 20:6b

Our God is in the heavens;
He does all that he pleases.
Psalm 115:3

Whatever the LORD pleases, he does,
in heaven and on earth,
in the seas and all deeps.
Psalm 135:6

I form light and create darkness,
I make well-being and create calamity,
I am the LORD, who does all these things.
Isaiah 45:7

For from him and through him and to him are all things.
Romans 11:36

And he is before all things, and in him all things hold together.
Colossians 1:17

God's word bellows His sovereignty. This fundamental truth underscores all subsequent assertions. Though one may easily declare God's sovereignty, reconciliation of this truth proves difficult. When we overlay the sovereignty of God upon life, actions, and particularly the will of men, conflict ensues almost immediately.

What about when we overlay the sovereignty of God on armed conflict? On war? On the bayoneting of a prisoner of war? On friendly fire?

I normally resist confessional declarations. However, men have historically codified beliefs concerning God into useful composed statements. The Westminster Confession of Faith, composed in 1646, stands as a timeless statement concerning orthodox biblical truth. It will assist us in a transition from this notion concerning sovereignty and authority toward *providence*, the outworking of God's sovereignty as seen in the actualization of the decree of God. Articles I and II of chapter 3 speak to providence,

---

Article I
God from all eternity, did, by the most wise and holy counsel of His own will, freely, and unchangeably ordain whatsoever comes to pass; yet so, as thereby neither is God the author of sin, nor is violence offered to the will of the creatures; nor is the liberty or contingency of second causes taken away, but rather established.[3]

Article II
Although God knows whatsoever may or can come to pass upon all supposed conditions; yet has He not decreed anything because He foresaw it as future, or as that which would come to pass upon such conditions.[4]

---

[3] "Westminster Confession of Faith," Center for Reformed Theology and Apologetics, accessed March 18, 2016, http://www.reformed.org/documents/wcf_with_proofs/.

[4] Ibid.

Providence

As we dive into the logical outworking of the sovereignty of God, the relationship between God and His creation, I'll increasingly seek assistance. Grudem defines providence in the following manner,

God is continually involved with all created things in such a way that he,

1. keeps them existing and maintaining the properties with which he created them;
2. cooperates with created things in every action, directing their distinctive properties to cause them to act as they do; and
3. directs them to fulfill his purposes.[5]

Grudem accurately assesses that this doctrine historically prompted substantial and significant disagreement among Christians "particularly with respect to God's relationship to the willing choices of moral creatures."[6] Though controversial and tough to digest, this doctrine contributes immeasurably to a useful and thorough exposition of the theology of war.

The Bible defines the decree of God. It tells us that God planned and ordained all things that come to pass. In eternity past, He decided what was to happen according to the counsel of His good will. God does not react to things. God does not respond to things. God directed every single thing that happened, is currently happening, and will ever happen. Observe David's confession,

in your book were written, every one of them,
the days that were formed for me,
when as yet there was none of them.
Psalm 139:16

His days, the days of David's life, were written beforehand, composed by the hand of the sovereign God. Job writes,

---

[5]Wayne Grudem, *Systematic Theology: An Introduction to Biblical Doctrine,* (Grand Rapids: Zondervan, 1994), 315.
[6]Ibid.

> *Since his days are determined,*
> *and the number of his months is with you,*
> *and you have appointed his limits that he cannot pass,*
> *Job 14:5*

Confirming the predestination of days, of life. Isaiah confirms,

> *declaring the end from the beginning*
> *and from ancient times things not yet done,*
> *saying, 'My counsel shall stand,*
> *and I will accomplish My purpose.'*
> *Isaiah 46:10*

God decreed His plan in eternity past, in "ancient times", and His plan shall stand, never to be thwarted. God will accomplish all that He proposes to accomplish. Paul verifies,

> *In him we have obtained an inheritance, having been predestined according to the purpose of him **who works all things according to the counsel of his will**, (emphasis mine)*
> *Ephesians 1:11*

He works all things to the counsel of His will, His perfectly good, holy, and righteous will. Only with much difficulty could one deny this fundamental truth regarding providence. In modernity, open theists not only deny aspects of providence and sovereignty but seek to systematize these errors. However, we will boldly and unashamedly declare what Scripture clearly delineates, that God's sovereign hand controls all that happens: earthquakes, fires and floods, traffic accidents and the like.

As it might be troubling to make these associations, an even greater difficulty arises in reconciling sovereignty with the presumed will of created, moral beings. The discussion always culminates on free will. Men quite naturally resist any notion of compulsion.

> But what about free will?
> I have free will, right?
> If you remove free will, does that make men robots?

As usual, we must carefully define terms. When most people say free will, they usually mean *libertarian* free will, meaning uncaused will free from external compulsion. With libertarian free will, I am free to make whatever moral decisions I decide based upon nothing whatsoever, uncompelled. The difficulties become intensely apparent when considered in light of salvation or in attempting to reconcile the evil actions of men, their wickedness, with the divine decree and the providence of God.

Free will exists only in that which is a slave to its nature. Libertarian free will, though coveted, exists only as a figment of the sinful imagination of men. All decisions, all choices, all actions are governed by something. The notion of a decision being made in an absolute vacuum is preposterous. A man's nature governs his will and all associated decisions as a function of his will.

Concurrence in Theory

How do we reconcile the actions of men, particularly the evil actions of men or even the necessarily violent actions of men participating in war, as directed by their hearts to the providence of God?

We must flesh out an idea concerning *cause*, and delineate primary from secondary causes. My will, whether sinful or not, drives my actions as a function of my nature. A greater cause overshadows my will and nature, a first cause. God is the primary cause as the two work in conjunction to bring about my actions. Though the Bible remains silent on *how* this occurs, it speaks clearly *that* it occurs. God directs the free will of men to serve His purposes. This is the crux of providence known as concurrence.

Grudem defines concurrence in the following manner:

> God cooperates with created things in every action, directing their distinctive properties to cause them to act as they do.[7]

Again, we join Nicodemus in exclaiming, "How can these things be?" Consider that the decree of God has ordered the steps of all men, yet men exercise free will in accordance with their nature entirely in line and supportive of the will and decree of God. Here is the nexus of the paradoxical aspects of God and His outworkings in history and the hearts of men.

As I promised, we won't dwell; we'll drive on with a parting shot. When asked to reconcile divine sovereignty with the notion of human

---

[7] Grudem, 317.

responsibility, Charles Spurgeon responded, "I wouldn't try. I never reconcile friends."[8]

Concurrence in Practice

What of evil and the problem of evil? How do we reconcile this with the providence of God and concurrence? We'll assess this issue from an individual and corporate aspect.

The book of Genesis records in the account of the patriarchs that the sons of Jacob sold their younger brother Joseph into slavery merely out of spite. They hated him and his arrogance. By the hand of God, Joseph rose to power in Egypt and eventually reunited with his brothers who beg his forgiveness upon the death of their father. Joseph responds, "As for you, you meant evil against me, God meant it for good." (Genesis 50:20) The brothers acted upon their wicked intent, to remove their pompous younger brother from their presence. Yet, God utilized their wicked intentions to accomplish His purposes. As we'll see, His purposes are always good.

The prophet Isaiah prophesied during a time when the infamously brutal Assyrian empire threatened imminent destruction upon Israel. In 722 B.C., God handed them over. Assyria destroyed Israel and enslaved many of its people. Speaking of Assyria, Isaiah writes,

> *Against a godless nation I send him,*
> *and against the people of my wrath*
> *I command him,*
> *to take spoil and seize plunder,*
> *and to tread them down like the*
> *mire of the streets.*
> *But he does not so intend,*
> *and his heart does not so think;*
> *but it is in his heart to destroy,*
> *and to cut off nations not a few;*
> *Isaiah 10:6,7*

What an astonishing passage! Who sent the Assyrians? The passage

---

[8]Robert L. Deffinbaugh, "Human Responsibility and Salvation," Bible.org, accessed March 17, 2016, https://bible.org/seriespage/12-human-responsibility-and-salvation-romans-930-1021.

clearly states that God did. He sent them to punish His people, the northern kingdom that had fallen into generations of apostasy. Did the Assyrians intend to serve as the rod of Yahweh's anger? No, they sought to satiate the wicked lusts of their own sinful hearts. In the spirit of concurrence, God directed their wicked wills to accomplish His divine purposes, in this case, judgment against Israel.

The will of God is perfect and good and always yields what is best.

And we know that for those who love God all things work together for good, for those who are called according to his purpose.
Romans 8:28

Paul makes the declarative statement that all things—not some things, but all things—work together for the good of those who are called according to God's purposes, speaking of God's people. Notice he doesn't declare all things good. However, he does declare that all things, good and bad, work together for the good of the believer.

Joseph's enslavement, clearly something we would declare as bad, worked together with other things for his good. The destruction of Israel at the hand of the Assyrians worked together with other things for the good of God's people. The wickedness and cruelty of the Assyrian regime puts this in a proper perspective, lends extreme *weight* to this idea. Again, we stand alongside Nicodemus pleading, "How can these things be?"

Consider the wickedest act in the history of mankind, the crucifixion of Jesus Christ. Jesus, as the Lamb of God, lived a perfectly sinless life, the only man in history to do so. He was unjustly arrested, unfairly tried and condemned, scourged, and delivered to the most painful of deaths on the cross at Calvary. At Pentecost, Peter stood to preach, declaring the words of the prophet Joel and then, speaking of the crucifixion, stated,

this Jesus, delivered up according to the definite plan and foreknowledge of God, you crucified and killed by the hands of lawless men.
Acts 2:23

Wicked men slew Jesus, the only innocent man in history, as a function of the darkness of their sinful hearts.

Yet it was the will of the LORD to crush him;
Isaiah 53:10

The crucifixion showcases concurrence at its greatest and most terrible, its fullest application. Lawless men slew Jesus. The decree of God prevailed in the slaying of Jesus. In sovereignty, His providence had already declared that this would be so, that this would be the means by which He would redeem His people.

> Who killed Jesus?
> Who hardened Pharaoh's heart?
> Who suffered Joseph in slavery?
> Who directed the destruction of Israel?

The answers to these questions yield the primary conclusions necessary to develop an active and accurate theology of war. As we turn to corporate bloodshed, we must reconcile the reality of God's intervention in human history, His providence.

As we recall the rich history of human bloodshed, consider this.

As we delve into the mind of the warrior, consider this.

As we dissect the warrior spirit, dwell upon this.

As we turn to matters of eternity, meditate upon this reality.

As we examine the work of God and the nations, ponder this.

My college major, engineering physics, along with my limited cognitive capacity drove me to be a great pragmatist. I could understand *what* very well. When presented with a problem to solve I could quickly ascertain the correct process, the correct equation, the correct method, yet I could rarely understand *why*. In many cases, I scored an 'A' on a test but could barely explain the underlying concepts being tested.

In much the same way, we'll declare the irreconcilable as true and proceed as, "The secret things belong to the Lord our God." (Deuteronomy 29:29)

Prepare to consider the implications of this truth. Know that these truths have challenged the greatest minds in history since they were established so long ago. I assure you, they are as necessary as they are difficult. As we seek deeper into the reality of war, allow these truths to transform your mind. Consider what you believe and compare to these declared truths. Examine them for what they are. Look to the masters. Look to the word of God.

Let us now look to the bloodshed.

Section 1: War and the Body

Chapter 4

<u>Friday, March 8th</u>

Again, two middle-aged men sat discussing a secret war against ISIS, this time at a school concert.

I'd been on leave for a couple of weeks when I ran into Mike again at the church where our kids attend school. As I helped my wife cut and glue white triangles onto black squares for a geometry display, Mike walked up holding his young son.

"What's up man, you ready?"

I assured him that I would be ready after I came off leave in a week. I intended to sign in, get an updated read-on, acquire some new kit, hit the range, and then catch a hop a week or so later. Painless.

"The boys got shot up the other night," he informed me.

My heart rate elevated slightly as I inquired of the circumstances. They had been on a daytime assault across the Forward Line of Troops (FLOT) to bad-guy land and had received quite bit of fire. On the upside, no one got hit, the aircraft were only slightly damaged, and they got to return fire. I know pilots who used to practically beg the bad guys to reveal themselves by engaging so they could shoot back. The downside is that they were fairly deep inside ISIS territory and had a round or two found their mark, things could've gotten ugly quick. Mildly, he informed me that they were relooking their procedures.

I thought ahead to this rotation and had a disturbing vision, being taken by ISIS. I envisioned the chaos and confusion of an engagement, the elevated radio chatter, the fog of battle. Often, especially during the daytime in a helicopter, it is exceptionally difficult to even tell if you've been engaged. On more than one occasion, post-flight inspections revealed bullet holes when the crew wasn't even aware of being shot at. I'd seen a single AK-47 round bring an aircraft down. Lucky shot or not, helicopters could be surprisingly tough or exceptionally fragile.

During Operation Red Wings in Afghanistan in 2005, Major Stephen Reich led a quick reaction force into a remote valley to rescue a stranded SEAL team. On short final, a single RPG round found its home and brought down the entire aircraft killing all 16 soldiers on board. Again, whether lucky or not, the effect is the same.

All aviators and the savviest of ground operators understand the fragility of these iron chariots and still they ride them into battle every single day, oftentimes throwing caution to the wind and rolling in hot despite the threat. I've always marveled that the bad guys didn't shoot down more helos than they do.

I envisioned the engagement. I felt the sickening drop in my gut, the high-pitched scream of engines being asked to do more than they are able, the wrestle for control, the call, "brace, brace, brace!", the bone-jarring impact, blackness and dust, the fight to sort it out, and then being taken.

I saw the rough hands and boots of the Daesh fighters. The flurry of sound and emotions, shouts of "Allahu Akhbar" ringing in my ears and I made a vow. At that point in time, I made a vow: I would fight to the death.

ISIS was becoming infamous for their brutality and had developed a surprising media savvy. Their media wing stages every aspect of their frequent executions for maximum shock value. They've become highly-produced events though that wasn't always the case. I was in Iraq in 2004 when Zarqawi beheaded Nick Berg.

Nick Berg worked as a contract civilian, a radio tower repairman, and in 2004, militants abducted him near Baghdad. Interestingly, his last known contact was with the liberal filmmaker Michael Moore who interviewed him for his anti-war film, Fahrenheit 9/11. Staying in the Al-Fanar hotel, Berg last contacted U.S. officials on April 10. His family never heard from him again until his decapitated body was discovered on a Baghdad overpass on May 8. They were told that his body showed "signs of trauma". Three days later, the enemy released the video.

By now, over a decade later, this new genre has become tiresome, so callous have our eyes and hearts become. The Nick Berg video broke new ground and at the time, shocked the world. In the footage, which was surprisingly simple and effective, Berg sits in an orange jumpsuit surrounded by five masked jihadis. This was at the height of the Abu Ghraib scandal whereby American soldiers were photographed mistreating Iraqi detainees. The orange jumpsuit was the uniform for detainees in Iraq and more notably at the U.S. detention facility in Guantanamo Bay. Cladding

Berg in a similar jumpsuit sent an overt message.

One of the militants reads a prepared statement, basically a laundry list of grievances, and when done he pulls a knife from his belt. The men pin Berg to the ground and unceremoniously saw off his head while chanting the *Takbir*, "Allahu Akbar" or "God is Great". At the conclusion, the leader brandishes the head while the others continue to chant.

Simple, grim, enormously effective—the video shocked the world, revealing the savagery and depravity of the enemy. The mainstream media edited out the barbarous portions of the clip, but I was stationed downtown with the boys at the time, and one asked me if I wanted to see it. I don't know what compelled me, but I said yes. I remember watching the video and hearing Nick Berg's screams and as they started to saw on his neck, something shifted inside of me, something internal. I looked away and turned it off, not wanting to see any more.

As of this writing, filmed executions have become extremely widespread among our Islamic enemies and though they have become more complex, slick and polished in comparison to the grainy Zarqawi video, the thematic elements and intended effects are still the same. ISIS has expanded their repertoire to include mass executions, shocking the viewer by the sheer volume, burnings, and other torturous deaths, but beheadings are still the main event.

I sometimes try to imagine what it must be like, kneeling on the sand in your orange jumpsuit under the noon-day heat, cameras rolling, your executioner standing next to you reading his statement and thinking to yourself, "I can't believe I am really about to get my head cut off."

One thing that shocks me is the apparent willingness with which the victims approach their death. As American photojournalist James Foley knelt at his hour of death, reading a list of grievances before Mohammed Emwazi beheads him, he exhibits a few obvious signs of distress, but otherwise remains calm, docile even. It is later revealed that he had undergone several mock executions prior to the actual event. Did this condition him to remain calm?

After ISIS overran Tikrit in 2014, they published videos of mass executions of hundreds of Iraqi soldiers and air force cadets. The videos showed the soldiers being hauled in flatbed trucks like cattle, herded into lines and marched to their demise. Handlers even issued derisive slaps on their rumps, herding them, moving them along. I wondered, why did no one go for a gun? Why did no one go down fighting, at least try to take one with you? They truly seemed like sheep before the slaughter.

I wondered if anyone ever spit in the face of their captors as they came brandishing the knife. I wondered if anyone ever stood in defiance while sporting the orange jumpsuit, refusing to go down kneeling. As the videos are highly edited, they would remove this footage, but did anyone die a good death?

In March of 2016, reports surfaced of an isolated Russian soldier fighting Daesh in Syria who, instead of surrendering, called in an airstrike on his own position, killing himself but taking with him a large contingent of enemy fighters. Would I have the fortitude to do the same?

As Mike and I chatted while trying to keep our kids in line, my mind raced through all of these scenarios. Were I not a religious man, I'd worry that I had jinxed myself by writing about my demise prior to it happening. Wouldn't that be epic? However, I know that God has numbered my steps, ordered my days, and as such, I'll do my best to honor Him as I live. I'll leave the rest up to providence.

4. The Source of War

The Schoolmasters

Could I ever do the schoolmasters justice? We possess neither the time nor the necessity to consider their vast and varied theories with adequate veracity. However, I desire to honor their labor.

I take care in not impugning their timeless work. The French general Jomini (1779-1869) postulated theories that became a staple at the United States Military Academy at West Point. Civil War generals, many of whom learned at West Point, employed Jominian tactics during the war though American terrain necessitated various adjustments. Even today, Jominian thought undergirds military instruction at academies throughout the west.[9]

As another poignant example of the continued relevance of the schoolmasters, Sun-Tzu writes,

> No country has ever profited from protracted warfare.[10]

He describes the impacts of extended warfare: the dulling of weapons,

---

[9] John Keegan, *The American Civil War*, (New York: Knopf, 2009), 96-97.
[10] Sun-Tzu, *The Art of War*, (New York: Barnes and Noble, 1994), 173.

depression of spirit, depletion of strength, and consumption of resources. As that happens, expect the enemy at the gates. As he puts it, "Feudal lords will take advantage of our exhaustion to arise."[11]

We see exactly this today. After 16 years, the United States continues to slog forward against the Taliban in Afghanistan. Already the longest war in American history and with no end in sight, the administration recently sought to increase troop levels after years of reductions. The U.S. withdrew from Iraq in 2010 after seven years of hard fighting. However, due to Islamic State resurgence and the collapse of the Iraqi military, the U.S. dispatched troops back to the region, mostly in an advise-and-assist role with the Iraqi military and the Kurdish Peshmerga. Recent indications suggest an imminent future deployment of combat troops.

Our enemies, taking a page from Sun-Tzu's playbook, pursue a strategy of attrition. A 2010 edition of *Inspire* magazine, the Al-Qaeda affiliated publication, celebrated the "strategy of a thousand cuts. The aim is to bleed the enemy to death."[12] Osama bin Laden acknowledged the strategy, that they could never defeat the enemy (America) directly on the battlefield. He believed that asymmetric attacks would effectively, over time, topple the economic backbone of the United States, which bin Laden saw as the American center-of-gravity. Indeed, a terrorist attack that might cost $10,000 might paralyze a nation and generate millions of dollars of economic fallout.

The outcome is still in doubt. Al-Qaeda, ISIS, and other like-minded groups deliberately seek to keep the United States engaged in a continual and protracted state of warfare, a tenuous condition per Sun-Tzu.

Sun-Tzu further demands that commanders and leaders run continual estimates and continually study the thing that is war. Sun-Tzu declares, "warfare is the greatest affair of state, the basis of life and death, the Way (Tao) to survival or extinction" and as such, "It (war) must be thoroughly pondered and analyzed."[13] The most important consideration of a state that prepares for war is the Tao (way). Nations must develop and promote an ideology, belief, or general feeling that unites people behind the ruler. In other words, "they will die with him; they will live with him and not fear

---

[11] Ibid.

[12] Matthew Cole, "Al Qaeda Promises U.S. Death By A 'Thousand Cuts'", *ABC News*, accessed March 29, 2016, http://abcnews.go.com/Blotter/al-qaeda-promises-us-death-thousand-cuts/story?id=12204726.

[13] Sun-Tzu, 167.

danger."[14]

Likewise, the Prussian Clausewitz posited intensely pragmatic insight noting that, "war is nothing but a duel on an extensive scale."[15] He viewed state conflict as akin to personal conflict on a macro level, the necessary friction of ideas, motives, and wills. He defined war in the following manner:

> War, therefore, is an act of violence intended to compel our opponent to fulfill our will.[16]

Bending your adversary to your will is the objective of not just war, but politics, which serves as the vehicle for the promotion of the nation's collective will. Harmonizing these two notions, war and politics, yields Clausewitz's oft-cited conclusion,

> We see, therefore, that War is not merely a political act, but also a real political instrument, a continuation of political commerce, a carrying out of the same by other means.[17]

As national objectives increasingly conflict with those of other nations, war becomes the means by which a nation pursues those objectives past the point at which peaceful coexistence of national objectives between states is possible. War is the furthest extreme along the continuum of *means* in achieving national political objectives.

Great truth exists within this insight yet I wonder if the great European wars of the 17th and 18th century colored Clausewitz's conclusions with a decidedly western slant. Had he engaged ISIS, who seek to usher the world toward the apocalypse, would he draw similar conclusions? What about when religious motivations replace or usurp anything political? What about when religious ideology generates action that secular logic would deem unreasonable? Can one-sided political objectives still yield favorable conclusions per his analysis?

Regardless, great utility exists in studying, scrutinizing, assessing and deliberating over the conclusions of Clausewitz, Sun-Tzu, Jomini, Douhet, Nye, and others. Surely their conclusions, which have endured for centuries,

---

[14] Ibid.

[15] Carl Von Clausewitz, *On War*, ed. Anatol Rapoport (London: Penguin Books, 1968), 101.

[16] Ibid.

[17] Ibid., 119.

demand inspection. As useful as they may be, they fall short with one regard, the notion of *origins*. Pragmatically, they provide great insight, operationally, tactically and even strategically, but what about an understanding of sources? From where does warfare actually derive? What actually generates systematic, corporate human violence?

Sun-Tzu notes,

> Thus what (motivates men) to slay the enemy is anger; what (stimulates them) to seize profits from the enemy is material goods.[18]

Clausewitz agrees, expounding upon the hostility and hatred that generate conflict and warfare.

> In short, even the most civilized nations may burn with passionate hatred of each other.[19]

Hatred proves useful in warfare. Nations and their militaries throughout history denigrated the enemy in varying ways to foment the hatred of the individual soldier for the enemy, thus increasing the likelihood that he will kill. We must account for national hatred in understanding war. However, none seem to peel the onion further. From where does this hatred originate? Why do nations hate? Why does this hatred predispose a nation to corporate violence?

We seek a deeper source. We seek not pragmatics, but rather insight into the true nature of war that it serves. From this, we'll inevitably develop pragmatics as our understanding of *why* will quite naturally drive us to *what* or *how*. Only then may we declare with any assurance a level of comprehensiveness.

Preliminaries

The aggregate of individual wills nurtures the collective ethos and accompanying corporate action. With national level sin, what we see is the general outworking of the summation of the individual hearts of men. The conduct of a nation normally reflects the hearts of its individual citizens.

Many factors may twist the aggregate into a less accurate reflection of

---

[18]Sun-tzu, 174.

[19]Ibid., 103.

the individuals who comprise the corporate, amplifying the position of some while suppressing that of others. State-level actors, corruption and coercion, the supernatural (not to be discounted): all may sway collective national will away from popular individual will. Consider that at its zenith a mere 10% of Germans professed allegiance to the Nazi party.[20]

The reciprocal proves equally true. Collective ethos cultivates individual attitudes. Consider that despite only 10% of Germans professing Nazism, many more *acted* like Nazis for a multitude of varying reasons.

In the relationship between individual and national level will, the prevailing winds of one drives the other but itself serves to be driven.

Therefore, to understand a collective, national activity such as war, we must first journey into the hearts of individual men. As with all things, we must return to the Garden where it all began to truly understand primary causes.

A Walk in the Garden

The entirety of Scripture hinges upon Genesis chapter Three. Absent this chapter the rest of the Bible succumbs to irrelevancy. Indeed, the Bible leans heavily upon the entire book of Genesis, but the central message of the Bible, the Gospel of Jesus Christ, depends entirely upon the events recorded in Genesis Three. The Gospel withers and dies apart from it. Removal or depreciation effectively destroys the theology of the cross before it's even written or declared.

Genesis chapters One and Two record that after creating everything—the light and dark, sun and moon, the earth and sky, the beasts of the field, the birds of the air, the fish in the sea—God said, "Let us make man in our image, after our likeness." (Genesis 1:26) Then, "the LORD God formed the man of dust from the ground and breathed into his nostrils the breath of life, and the man became a living creature." (Genesis 2:7)

A creature in the image of God walked the earth for the first time, possessing the literal breath of God, the very breath of life. All other things God spoke into existence, but not man. He formed man with His hands. God created a garden called Eden and placed the man in it, to work it, to keep it, to have dominion over it. He issued the man a command,

---

[20] Chris McNab, *Hitler's Master Plan*, (Amber Books, 2011), 22-23.

You may surely eat of every tree of the garden, but of the tree of knowledge of good and evil you shall not eat, for in the day that you eat of it you shall surely die.
Genesis 2:16-17

Then, as man worked the Garden, God remarked, "It is not good that the man should be alone." (2:18) He brings every creature before the man to be named but finds no suitable helper so God causes the man to sleep. He removes a rib from his side and forms woman and presents her to the man. His epic response, upon being confronted with the sheer reality of woman, never having previously beheld or even conceived of this special being:

This, at last, is bone of my bones
and flesh of my flesh;
she shall be called Woman,
because she was taken out of Man.
Genesis 2:23

Scripture then says that, "Therefore a man shall leave his father and his mother and hold fast to his wife, and they shall become one flesh." (v. 24) They shall literally cleave to one another, a mingling of souls, forsaking all others.

The two lived in the Garden and enjoyed perfect harmony. Every relationship resonated with harmony: between man and woman, man and creation, and most importantly between man and God. Adam and Eve lived in the presence of God, basking in fellowship with the Author of all things. This brings us to the most critical book of the Bible, Genesis Three, which is the hinge-point of history. Genesis Three records the Fall of man, an event with eternal implications.

Scripture records that Satan roamed the garden as a serpent and was craftier than any other beasts of the field. (Genesis 3:1, Revelation 20:2) One day, the serpent came to Eve and introduced doubt into her mind about what God had commanded. He corrupted the words of God and deceived her into eating the fruit from the tree of the knowledge of good and evil. Eventually, Adam also consumes the fruit as Eve gives it to him.

Instantly, innocence is lost in their disobedience and they hide from God realizing they are naked for the first time. God confronts them and inquires as to what happened as if He weren't already aware. Both Adam and Eve defer guilt.

"The woman whom you gave to be with me, she gave me fruit of the tree, and I ate." (v. 3:12)

"The serpent deceived me, and I ate." (v.13)

Both the woman and the man deferred guilt to the other, to the serpent, and even to God. Notice the man's declaration, "the woman *you gave to me* (emphasis mine)". He deliberately caveats the declaration, reminding God that He, in fact, gave the woman to him, a far cry from his first declaration concerning the woman, "Here at last!"

The rest of chapter three records God's address to all three malefactors: Satan, Eve and Adam. The imminent effects of His words resonate throughout history, indeed, throughout the heart of every man that has ever lived.

Adam and Eve's betrayal shattered harmony. Discord, enmity, and separation supplanted the previous harmony between man and creation, man and woman, and most of all, between man and God. With the commission of the first sin, Adam corrupted men for all eternity. "Sin came into the world through one man, and death through sin," (Romans 5:12) leading to condemnation for all men. From the Fall, man declared himself an enemy of God, a rebel against the Creator, thus dooming himself to eternal separation from God.

The Fall corrupted all of creation, spoiling it. Paul writes,

> *For the creation was **subjected to futility**, not willingly but because of him who subjected it, in hope that the creation itself will be set free from its **bondage to corruption** and obtain the freedom of the glory of the children of God. For we know that the whole creation has been **groaning** together in the pains of childbirth until now.*
> *Romans 8:20-22*

Notice that creation is subjected to futility. It is in bondage to corruption. Harmony ruled. Now, corruption dominates. Sin corrupted all things. All things exist as mere shadows of their former selves. Creation groans in agony under the weight of bondage, corruption, and futility.

Brave Warrior, you may wonder what all of this could possibly have to do with the conduct of war. The humble answer is, everything! To fully understand war, to truly understand primary causes, to fully delineate the temporal from the effectual, we must fully define and understand the context of reality. The broadest foundational grasp of the basis of reality will yield our greatest chance to accurately define the reality of war.

Cursed Existence

Following the confrontation, God issues a series of proclamations (curses).

We'll deliberately bypass Genesis 3:15, God's address to Satan, known as the *protoevangelium* whereby God first declared the Gospel message to Satan himself, destining Satan for inevitable defeat. We know now, through the prism of the New Testament, that Christ on the cross actualized Satan's defeat.

Instead, as we dive into the hearts of men to understand the thing that is war, we'll begin by examining God's curse upon Eve and the man, Adam.

To the woman, God declares that childbirth would now be an intensely painful process. No longer would she bear children with ease. As I've heard that childbirth could be painful, I'll take this as substantive! The second half of God's curse upon her demands a more thorough examination. God says to her,

> *Your desire shall be for your husband, and he shall rule over you.*
> Genesis 3:16b

On the surface, at least part of this curse sounds like an okay thing. Her desire shall be for her husband. That actually sounds good and healthy. Nothing wrong with a wife desiring her husband. The language betrays this notion. The word used for desire is *tĕshuwqah* from the Hebrew, a noun expressing desire with connotations of consumption, or a consuming. The very next chapter in Genesis will assist us in ascertaining what this curse actually means.

God, speaking to Cain concerning his ever-present sin, tells him that it (sin) is crouching at his door and that "its *desire* is for you, but you must rule over it." (Genesis 4:7b) Sin's *desire* is for Cain. Cain's only authority should be God but sin seeks to usurp authority in his life, to consume him. God tells Cain that he must rule over his sin implying that his sin is attempting to rule over him. Its *desire* is for him.

Returning to the garden, God tells Eve that her desire shall be for her husband. She shall seek to rule over him, to usurp authority in his life. A biblical understanding of complementarianism provides a broader understanding of the problematic nature of this declaration. We'll explore this more fully in Chapter 27 but for now, observe the requisite friction between first half of the curse on the woman with the second, "and he shall rule over you."

The woman will seek to subjugate the man, but the man, in fact, will rule, dominate, exercise ungodly dominion over the woman. The conflict is obvious. The enmity between man and woman profoundly impacts every single human to ever walk the face of the earth.

What of Adam's curse? God says to the man,

cursed is the ground because of you;
in pain you shall eat of it all the days
Of your life;
thorns and thistles it shall bring forth
for you;
And you shall eat of the plants of the field.
By the sweat of your face
you shall eat bread,
till you return to the ground,
For out of it you were taken;
for you are dust,
and to dust you shall return.

Be not mistaken, the primary and most significant consequence of the Fall and subsequent curse is death, instant spiritual death followed by physical death. As a result of the Fall, all men are born spiritually dead, which is why we require new birth to live. The broken relationship with God yields eternal consequences. The other temporal aspects bear discussion as they generate numerous tangible ramifications.

Genesis Two states that God made man and put him in the Garden "to work it and to keep it." (v. 15) The author uses precisely the same language Moses uses in Numbers Chapters Three and Four as he describes the work the Levites do in caring for God's tabernacle. This is not just work for the purpose of work, rather work as a ministry to the Lord. In Adam's work in keeping the Garden, he ministers to God. This is a holy vocation. The call to work is a holy call.

In its perfection, creation initially yielded all that man needed and desired. Sustainment required no effort on behalf of man. The curse effectively destroyed this condition. Man would still eat. Man would still live, yet only by struggle. Creation will now yield all that man needs, unwillingly. Only by "the sweat of his brow" would man meet his requirements and desires. Work, still a holy vocation, now becomes a struggle. As much as the curse changed creation, it likewise changed the desires of men.

Several millennia later the Apostle Paul, in condemning the wicked and unrighteous, declares that man "exchanged the truth about God for a lie and worshiped and served the creature rather than the Creator." (Romans 1:25) Man, in effect, declared himself the captain of his own destiny, the master of his own fate, and no longer worshiped the Creator. Instead, man now worshiped the created thing, the creature.

Consider that which the overwhelming majority of men pursue. John defines the objects of this pursuit as "the desires of the flesh, and the desires of the eyes and pride of life." (1 John 2:16). After the Fall, the sheer act of obtaining and acquiring consumes men. No longer do men pursue that which is honorable to God, that which is freely given. Men now worship the created thing alongside themselves and yet, the created thing exists as a mere shadow of its prior glory, just as they do.

Corrupted creation and corrupted desires collude to frustrate the desires of all men. Life withholds that which men now desire. Because of the corruption of creation at the Fall, working to obtain has become exceedingly difficult. You could say that with this curse, God enslaved men to the rat race, the daily struggle for sustenance and acquisition.

An analogy may prove useful. Consider a car accident. I've seen a vehicle after a head-on collision with a tractor-trailer, horribly twisted by the sheer awful physics of the accident but even after the collision, still recognizable as a car. Now, the accident distorted and twisted it into something much different than the original creator intended, but I still recognize the original design. Sin twists men and distorts created things though I can still recognize them for what they are by their original design. In their distortion, they exist as a parody of their former selves, not what the Creator had in mind, ghosts of a lost image.

Still, men seek fulfillment in these twisted and distorted created things, things that will never fulfill. Men tear themselves apart acquiring this destroyed car. Maybe it's been in a minor fender bender. Maybe it's been rolled, totaled even. Men strive for this. They labor, toil, and maybe even obtain. They cannot drive it and no matter how hard they try, how skilled they may be, they can never twist it back into its original shape. They can never regain the glory of that which is lost and their soul grieves this loss, paradise lost, even if they cannot articulate or even understand the empty anguish, the hollow nature of their pursuit. They've never actually seen the untouched car, they just come to realize that what they do end up possessing somehow falls short.

This is the darkness common to the struggle of all men and it permeates every aspect of existence. Announced by God in the Garden, the curse quickly registered its presence.

Cursed Action

Within the immediate generation following Adam and Eve, sin manifested itself in the first recorded act of violence. Their son Cain struck down Abel in a fit of rage. Prior to the Fall, there was no murder and no death in fact. This singular act announced the Curse to creation.

Cain, as the firstborn son of Adam and Eve, should have maintained a place of honor and respect. He exchanged that on the basis of anger at a perceived slight regarding his faithlessness. The brothers, Cain and Abel, brought their offerings before the Lord. Abel, a shepherd, brought the firstborn of his flock. Scripture is silent on what Cain, a worker of the ground, brought only that "the LORD had regard for Abel and his offering, but for Cain and his offering he had no regard." (Genesis 4:4,5) Cain's self-righteous anger spills over into a petulant exchange with God.

Afterward, he encounters his brother in the field and he "rose up against his brother Abel and killed him." (v. 8) John later speaks to Cain's sin, admonishing believers that, "We should not be like Cain, who was of the evil one and murdered his brother. And why did he murder him? Because his own deeds were evil and his brother's righteous." (1 John 3:12) We discern that Abel presented a righteous offering to the Lord while Cain did not. Perhaps Cain's sacrifice was damaged or minimal, maybe no sacrifice at all. Likely, he did not present his best as God demands, and for this God had no regard.

Consider that Abel's sacrifice had no direct bearing on Cain. The LORD had no regard for Cain's sacrifice because of what it was, not because Abel's was better. Had he presented an acceptable offering, God would have given it regard. Abel bore no responsibility for the rejection of Cain's offering, yet Cain's anger so burned against Abel that he raised his hand against him.

Today, we're mostly fine with murder. People get murdered every day. We watch murder on television and in movies. We read about murder in the news. Video games make a sport of murder. By their teen years, children have witnessed thousands of murders, digital and otherwise. Murder just fails to shock anymore as we've become callous to the taking of human life.

Consider that Cain never saw a murder. Cain never heard of murder. How did he know to even take life? John says that he was "of the evil one", implying that perhaps Satan suggested it.

Consider Cain standing in the field, contemplating his unwitting brother, his very flesh and blood.

"Strike him down," whispers Satan.

"It's his fault."

"He's trying to make you look bad."

"He has what you don't."

"He thinks he is better than you."

At some point, the voice of Satan becomes indistinguishable from the voice of the flesh and Cain picks up whatever and strikes down his brother.

James, the Lord's half-brother, attributes outward dysfunctional expressions to an internal struggle. In describing wisdom, he declares that "if you have bitter jealousy and selfish ambition in your hearts…this is not the wisdom that comes down from above." (James 3:14,15) He further describes jealousy and selfish ambition as earthly, unspiritual and yes, demonic. Wherever jealousy and selfish ambition govern the hearts of men, disorder and chaos reigns. James calls it "every vile practice." (v.16)

Physical violence against our fellow man stands in opposition to God's original, *revealed* plan. The demonic attributes of jealousy and selfish ambition drive men to violence against their fellow men, a most vile practice. James asks,

> *What causes quarrels and what causes fights among you? Is it not this, that your passions are at war within you? You desire and do not have, so you murder. You covet and cannot obtain, so you fight and quarrel.*
> *James 4:1-2a*

Clearly, in the heart of Cain, passions warred with one another. Surely he loved his brother Abel at some point or in some way, though Scripture is silent on the issue. Jealousy and selfish ambition warred within him, passions driving him from the rock of what he knew to be right. He clearly desired acceptance from God, just not at the expense of a suitable offering. Therefore, he murdered. He coveted his brother's position before God, thus he slew him.

Sin now had legs in the form of physical violence. The wickedness in the human heart had crowned a champion. Murder, physical violence against a human, constituted an actual assault against the very Image of God; sin in its fullest expression if you will.

Cursed Incorporation

Though by His very definition, God is impassible, not governed by emotions, various encounters seem to indicate the heart of God as prone to great and vivid passions. As man and sin spread across the planet, wickedness multiplied greatly and God actually regretted making man. Scripture records that "it grieved him to his heart." (Genesis 6:6b) God determined that He must remove the blight of man from the face of the earth, declaring but eight worthy of salvation, Noah, his three sons, and their wives.

Four specific transgressions doom all others:
1. Wickedness (Genesis 6:5a)
2. Continuous evil residing in the hearts of men (Genesis 6:5b)
3. Corruption of all things (Genesis 6:11a)
4. Violence (Genesis 6:11b)

Violence, as the fullest expression of men's corrupted hearts, doomed the entire planet to destruction. Scripture remains silent on the nature of the violence. Was it corporate? Systematic? Genocidal? Was it a continuance, on a global scale, of the individual violence in the manner of Cain, an anarchy of sorts? Perhaps it was a combination. Either way, violence as a manifestation of sin grieved God to the point that He regretted making man and designated all for destruction.

After the flood, Genesis 13 records the first account of corporate, collective friction. God called Abraham from Ur to the land of the Canaanites for a specific future purpose. Abraham and his nephew Lot, both rich in livestock found that "the land could not support both of them." (Genesis 13:6)

> *there was strife between the herdsmen of Abraham's livestock and the herdsmen of Lot's livestock.*
> *Genesis 13:7*

Though they resided in Canaan, Abraham and Lot agreed to separate for land was plentiful. Though a peaceful solution prevailed, one can easily envision a minor scuffle between herdsmen threatening to erupt into a broader conflict similar to how the assassination of the Archduke Ferdinand spawned the First World War.

Genesis 14 provides the first account of an organized conflict that in a brief narrative spans the gamut of military operations from insurgency to counterinsurgency, to occupation, to high-intensity conflict, to a singular raid.

Chedorlaomer is the dominant king of Elam and after 12 years of subjugation, five kings of the Valley of Siddim rebel against his rule. King Bera of Sodom, Birsha of Gomorrah, Shinab of Admah, Shemember of Zeboiim and the king of Bela(Zoar) join forces in rebellion against the Elamites. They likely object to a requirement to pay tribute. Chedorlaomer and three vassal kings, Amraphel of Shiner, Arioch of Ellaser, and Tidal of Goiin initiate a campaign of occupation to quell the rebellion. Eventually, they subdue the entire area by force.

Inevitably, the two sides meet in a decisive engagement in the Valley of Siddim. Chedorlaomer and his allies thoroughly trounce the rebels, sending them fleeing into the hill country. As was the custom in those days, Chedorlaomer and his armies loot the cities left behind, plundering their possessions, enslaving their women. They also took Abraham's nephew Lot captive and departed for Dan.

Abraham quickly mustered a fighting force of 318 trained men from his allies and set out in pursuit. He overtook Lot's captors at Dan and executed the first recorded night raid, dividing his forces and flanking the enemy. So thorough was the rout that he not only rescued Lot and all of the possessions, but he pursued the enemy all the way to the north of Damascus before returning in triumph.

In this first account of corporate violence, we see the culmination of the progression of sin from the Fall whereby sin entered the world, to the Curse, to the first recorded account of individual violence, and to the military campaign at Siddim, the first account of corporate violence. This first account displays of its own regard the entire spectrum of violence as a function of sin. In this, it clarifies what the schoolmasters lacked, a true understanding of origins which is necessary if we are to seek a better understanding.

A Partial Definition

Our study of origins allows us to partially define war. As the schoolmasters failed to account for the supernatural in looking to the source, their attempts at defining war fall precipitously short and remain inconclusive.

In beginning to understand the thing that is war, the Fall's impact and bearing upon our present existence drives us to the following *incomplete* conclusion:

> *War is the systematic corporate friction between the collective wills of sinful men...*

Only when we overlay the reality of God upon the bloodshed may we fully define war. For that, let us return to the invasion and then assess it in terms of the sovereignty of God.

5. Road to War

Upon the Illumination of the Spirit

Life exists as a series of revelations. The unregenerate man lives blinded by sin and flesh to the truth of the Gospel message. Then, like an explosion of life and reality, the Holy Spirit quickens the soul in the supernatural act of regeneration. The man formerly blind, may now see. One of my favorite testimonies concerns an Army friend of mine and this very thing.

While deployed to Afghanistan he spent night after night with a certain friend of his, poring over Scripture, going through verses of the Bible. His friend still walked in blindness, stumbling toward the slaughter. (Pr. 24:11) He just didn't get it. It didn't make sense to him which ironically makes sense as Scripture itself declares, "the word of the cross is folly to those who are perishing." (1 Corinthians 1:18) The Bible reeks of foolishness to the unregenerate man.

Night after night, my friend never gave up, pouring into his friend's heart. Then one evening, it happened. The man looked at my friend with a newness of eyes and said matter-of-factly, "I get it now. I understand. I'm saved." It was that simple, that undramatic. The Lord lifted the veil, calling him out of the darkness and into His marvelous light. He didn't need a ritual or a magic formula, a priest or a church, just the newness of sight in the formerly blind.

However, what appeared simple and undramatic in the flesh masked a spiritual melee, an epic spiritual struggle for the eternal soul of this man, a violent and intense dogfight, a desperate conflict with eternity hanging in the balance. On the surface, the man simply understood for the first time as the Lord revealed to Him his sinfulness and his need for an all-sufficient Savior. Upon revelation, he believed.

God's first revelation, saving knowledge, calls us forth from the darkness. Prior to this, our darkened minds perceive nothing. Once a man is saved, the Holy Spirit serves as his instructor, illuminating the truth of God's word through transformation and the renewal of the mind. (Romans 12:2) I remember God first ripping the veil from my own eyes concerning salvation—not without much pain I might add. I also remember the very first revelation of the Holy Spirit to me. They asked me to teach Sunday school.

Though I'd been raised in church, the Bible remained a mystery to me, especially the Old Testament. In the feebleness of my darkened mind, the Bible, but especially the Old Testament, existed as a collection of unrelated stories and fables that perhaps served a moral purpose. The notion of unity and cohesion never entered my mind. Thus, my view concerning Scripture amounted to a fuzzy collection of questionable accounts of lions' dens and giant fish. David and Goliath. Noah and the Ark. Samson and Delilah. I could even recall a song I sang as a child, *"Joshua fought the battle of Jericho, Jericho, Jericho."*

These existed in my mind as something akin to the Grimm fairy tales, a misunderstanding the Spirit seemed quick to dispel. After being saved and baptized, Ami and I joined our first Sunday school class. After a number of weeks, as we became fairly comfortable in our new surroundings, the instructor asked me to teach. He'd be out of town and so why not ask the brand new believer who knows absolutely nothing about the Bible to teach? I saw nothing unusual about this at the time.

Like any military man, when given a mission, I put my heart into it determined to make good on the trust placed in me. At the time, we were studying the Babylonian exile and as I studied and prepared, at some point a switch flipped, a light came on, a veil was lifted once more.

> Wait a minute, these things are all related!
> These are the same people from that previous account.
> There is a larger narrative, a broader purpose!

The sudden illumination of the Holy Spirit practically knocked me to the floor. I felt like Archimedes crying 'Eureka!' The pages of Scripture radiated truth unto my soul. The Holy Spirit taught me for the first time as I devoured Scripture with a newfound hunger. I consumed like a starving man presented with the most perfectly grilled and succulent T-bone steak imaginable.

That we would all consume Scripture with unwavering veracity. Truth and salvation reside within. Seek it; pray that God would give you a hunger for His word. Pray that the Holy Spirit would illuminate the infinite truths of the Bible, that He would quicken your mind to grasp the deepest of things. As the Bible testifies about itself,

> *For the word of God is living and active, sharper than any two-edged sword, piercing to the division of soul and of spirit, of joints and of marrow, and discerning the thoughts and intentions of the heart.*
> *Hebrews 4:12*

The word of God pierces and divides, transforms and renews. It is not for the feeble or weak of heart. Realize that pursuing the truths of the Bible places your very existence at risk. Are you prepared for that?

Oh, and yes, my Sunday school class went very well that next week.

About the Road to War

It all began with a promise. God sought Abraham and commanded, "Go…to the land that I will show you." (Genesis 12:1) Consider the reality of that order, the implications. God called Abraham to "Go" without telling him exactly where to go. He then tells Abraham that He will make him into a great nation and will bless all the families of the earth through him. (vs. 2,3) Abraham believed God and went. He departed Ur and Haran, modern day Iraq, and traveled west to Canaan, modern day Israel.

I spent a lot of time in western Iraq and I distinctly remember flying up and down the Euphrates River from Baghdad, west over Habbaniyah and Ramadi, and out to Al Qaim on the Syrian border. I recall the infinite expanse of bright and rolling sand interrupted by the deep green ribbon of the ancient Euphrates. Palm groves line its banks hiding the scattered and random mud compound. It exists today, I'm sure, much as it existed in the days of Abraham, and from the air, it always struck me that Abraham walked this land. He drank these waters. He rested on its banks when weary. The reality of Abraham's call, when considered over the actual terrain of his call, always struck a chord within me.

Upon arrival in Canaan, God reaffirmed the covenant, His promise to Abraham.

> *Look toward the heaven, and number the stars, if you are able to number them…So shall your offspring be.*
> *Genesis 15:5*

> *I am the LORD who brought you out from Ur of the Chaldeans to give you this land to possess.*
> Genesis 15:7

> *On that day the LORD made a covenant with Abraham, saying, "To your offspring I give this land,...*
> Genesis 15:18a

> *and in your offspring shall all the nations of the earth be blessed,*
> Genesis 22:18

God promised to make Abraham into a great nation, to give him numerous offspring. He promised him the land that is Canaan and again, He promised to bless the earth through Abraham's offspring…just not yet.

There was a problem. People already occupied the land of Canaan. What about them? God's promise to Abraham was intertwined with the fate of the Canaanites (Amorites).

> *Know for certain that your offspring (Israel) will be sojourners in a land that is not theirs and will be servants there, and they will be afflicted for four hundred years…and they shall come back here in the fourth generation, for the iniquity of the Amorites is not yet complete.*
> Genesis 15:13,16

God, in His mercy, saw fit to delay fulfillment of his covenant with Abraham for 400 years, as "the iniquity of the Amorites (was) not yet complete." (v. 16) God allowed the Amorites, the people in Canaan, 400 additional years to repent and turn from their iniquity. God's mercy resonates in that He ordained His chosen people to suffer for 400 years "in a land that is not theirs" speaking to their impending and intended bondage in Egypt.

From Whence Canaan

Many fretter about God's later commands to Israel concerning the complete destruction of the Canaanites. A historical dive into their origins proves useful in putting this command into perspective.

Several millennia prior, after the flood resided, Noah and his sons emerged from the ark to find a changed world. Genesis Nine records God's covenant with Noah and Noah was so joyous, so overcome with gratitude to

God for sparing him and his family from judgment that he celebrated by...getting drunk.

As he passed out naked in his tent, his youngest son Ham peered in and then went and told his brothers, Shem and Japheth. They also came but covered up their father without looking at him in the shame of his nakedness. As Noah awoke and saw what his son Ham had done, gazing upon him as he lay naked and drunk, ostensibly speaking in jest to his brothers, Noah pronounced a curse upon him,

> *Cursed be Canaan;*
> *a servant of servants shall he be to his brothers.*
> *Blessed be the LORD, the God of Shem;*
> *and let Canaan be his servant.*
> *May God enlarge Japheth,*
> *and let him dwell in the tents of Shem,*
> *and let Canaan be his servants.*
> *Genesis 9:25-27*

This "Curse of Canaan" has baffled students of God's word for millennia. To Shem and Japheth, he declared that Canaan shall be their servant without explaining exactly what that means or why he cursed Ham's yet-to-be-conceived son and not Ham himself. Did God even support the curse? Noah declares it, but God never mentions it. The Table of Nations (Genesis 10) definitely lists Canaan as one of Ham's sons, along with Cush, Egypt, and Put. (v.6) The Table likewise records Canaan's offspring.

> *Canaan fathered Sidon his firstborn and Heth, and the Jebusites,*
> *the Amorites, the Girgashites, the Hivites, the Arkites, the*
> *Sinites, the Arvadites, the Zemarites, and the Hamathites.*
> *Genesis 10:15-18a*

Frequently, the Old Testament speaks of the '-ites' in various situations, usually referring to the people groups descended from Canaan that occupied modern day Israel, but not always.

The Moabites and Ammonites descended from Abraham's nephew Lot after his two daughters got him drunk and slept with him following the destruction of Sodom and Gomorrah. These nocuous origins contributed to Israeli hatred toward the Moabites and Ammonites. Israel viewed them as an impure or unclean race. Similarly, the Edomites descended from Esau, Jacob's brother, and they were likewise antagonistic toward Israel. All three groups—Edom, Ammon, and Moab—come from the line of Shem in the Table of Nations, the same line as Israel.

However, the rest of the '-ites' descended from Ham and were cursed to serve the others. Collectively, the Bible often refers to the '-ites' as Canaanites. Sometimes it uses the labels interchangeably e.g. Canaanite versus Amorite. Nevertheless, this people group inhabited the land now sworn to Abraham.

What of their iniquity that yielded destruction? God condemned the Canaanites for rampant idolatry, pagan worship, and forsaking God. Their judgment was certain.

> *...it is because of the wickedness of these nations that the LORD is driving them out before you.*
> *Deuteronomy 9:5b*

God later spelled out the charges in detail, urging His people not to follow "the abominable practices of those nations (Canaan)." (Deuteronomy 18:9) These abominable practices included burning their sons and daughters as offerings and practicing sorcery such as divination, fortune telling, and contacting the dead, all forms of the occult. (vs.10-12)

Centuries later, God condemns the wicked king Ahab because "he acted very abominably in going after idols, as the Amorites had done, whom the LORD cast out before the people of Israel." (1 Kings 21:25) Clearly, the Canaanites rejected God and deserved judgment. God in His mercy offered them a 400-year respite at the expense of His very own.

Of Bondage and Deliverance

The Patriarchs dominate the book of Genesis. The first few chapters describe Creation, the Fall, and the Flood but the bulk of the book chronicles the origins and early history of God's people. God's covenantal love unifies His interaction with Abraham, Isaac, and Jacob (Israel)—the Patriarchs of Israel.

Jacob's twelve sons subsequently spawned the twelve tribes of Israel. Through some dramatic circumstances replete with betrayal, intrigue, and deception, all under the providential hand of God, Jacob and his sons ended up in Egypt, fleeing the harsh famine in the land of Canaan. Genesis concludes with the deaths of Jacob (Israel) and his son Joseph, who had been instrumental in God's providential plan to move his intended people to Egypt.

Exodus opens with Israel prospering in Egypt.

> *But the people of Israel were fruitful and increased greatly; they multiplied and grew exceedingly strong, so that the land was filled with them.*
> *Exodus 1:7*

The previous administration in Egypt viewed Israel favorably as Joseph rose to a position of national prominence within the regime, second only to Pharaoh. Conditions soon changed.

> *Now there arose a new king over Egypt, who did not know Joseph.*
> *And he said to his people, "Behold, the people of Israel are too many and too mighty for us. Come, let us deal shrewdly with them, lest they multiply, and, if war breaks out, they join our enemies and fight against us and escape from the land."*
> *Therefore they set taskmasters over them to afflict them with heavy burdens.*
> *Exodus 1:8-11*

The new king feared this numerous minority in Egypt and so he "made the people of Israel work as slaves and made their lives bitter with hard service." (Ex. 1:13) Scripture records that they built "store cities" for Pharaoh who brutally oppressed them. The Egyptians themselves descended from Ham (Genesis 10) and likely brought writing and culture including pyramid building from Babel, the sight of the great ziggurat that so offended God. (Genesis 13)

An interesting line of query we'll indulge, not essential to this work but interesting nonetheless, is whether the Hebrew slaves built the pyramids. Much of the discussion centers around the differing dates of certain Egyptian dynasties with respect to the dates of the Israeli captivity. The differences seem fairly insubstantial as they are disputed and likewise concern a hundred years or two, here and there. One would expect some fluidity in establishing the dates of events that have been buried under time and sand for thousands of years.

Pyramid technology existed at that time and in oppression, Egypt enslaved the vast nation of Israel and "made their lives bitter with hard service, in mortar and brick, and in all kinds of work in the field." (Exodus 1:13) Did this include pyramid building?

Amenemhet III built the last great pyramid and is the potential foster father of Moses. His daughter, Sobekneferu, had no son to succeed her rule and perhaps she was the "daughter of Pharaoh (who) came down to bathe at the river" and found baby Moses in the basket. As an adult, Moses killed an Egyptian who he witnessed beating a Jew prompting his flight to Midian, only to return after 40 years at the call of God to free the Hebrew nation. (Exodus 3) Upon his return, Moses confronted Pharaoh, likely Neferhotep I, who eventually relented and allowed Israel to leave Egypt. Scripture records that he perished at the Red Sea leading the pursuit of Israel. (Exodus 14:27)

The Jewish historian Josephus recorded that, "They (the Egyptian taskmasters) set them also to build pyramids."[21] Other extra-biblical sources indicate the presence of Asiatic slaves in Egypt during the approximate time that at least some of the pyramids were built. In her book, *The Pyramid Builders of Ancient Egypt*, Dr. Rosalie David noted the presence of the slaves and their disappearance with seemingly no explanation.[22] She wrote, "There are different opinions of how this first period of occupation at Kahun drew to a close. ... The quantity, range and type of articles of everyday use which were left behind in the houses may indeed suggest that the departure was sudden and unpremeditated."[23] Further, researchers have never uncovered the tomb of Neferhotep I, which is in line with the biblical account of his death in the Red Sea.[24]

Whatever occurred during the captivity, the Jews clearly suffered during their enslavement in Egypt. They suffered at the hand of their own God as He extended mercy to the Amorites, a people who hated God. Meditate upon this truth and see the sovereign hand of God in the most unexpected events. The Exodus, though it saved the Jews from Pharaoh's hand, drove them to the wilderness (Sinai Peninsula) where they would wander for 40 years anticipating the fulfillment of a promise they would never see.

---

[21] Flavius Josephus, *Antiquities of the Jews,* II-IX-1 accessed April 4, 2016, http://www.biblestudytools.com/history/flavius-josephus/antiquities-jews/book-2/chapter-9.html.

[22] A.R. David, *The Pyramid Builders of Ancient Egypt: A modern investigation of Pharaoh's workforce*, (London: Guild Publishing, 1986), 191.

[23] Ibid., 199.

[24] Nicolas Grimal, *A History of Ancient Egypt*, (Wiley-Blackwell, 1994), 184.

A Promise Thwarted

Even the most committed believer will doubt at some point. Despite all that they had witnessed—the parting of the Red Sea, God Himself leading them in a pillar of fire, the issue of the Ten Commandments—Israel continuously flirted with disbelief. It would not be without consequences.

God brought them to the wilderness of Paran at Kadesh-Barnea where He commanded them, "Send men to spy out the land of Canaan." (Numbers 13:2) Moses selected men as he was commanded and deployed them to Canaan. After forty days, they returned and issued their report. The land was all that God said it would be,

It flows with milk and honey, and this is its fruit.
Numbers 13:27

They expressed concern with the occupants of the land seeing their strength and fortified cities. Thus they lamented,

The land...is a land that devours its inhabitants, and all the people that we saw in it are of great height.
Numbers 13:32

In light of this report, their spirit flagged.

We are not able to go up against the people, for they are stronger than we are.
Numbers 13:31

Only Caleb remained steadfast exclaiming, "Let us go up at once and occupy it." (v. 30) Otherwise, a uniform corporate lament overtook Israel as they "raised a loud cry" and "wept that night" and "all the people of Israel grumbled against Moses and Aaron." (Numbers 14:1,2) Caleb and Joshua both tried to encourage Israel, as they had been among the spies, but the people responded by trying to stone them to death.

Only the glory of the LORD appearing at the tent of meeting saved them. God's anger at the people's lack of faith burned as He declared their judgment to Moses. Moses interceded and God relented somewhat, declaring,

...none of the men...shall see the land that I swore to give to their fathers...
Numbers 14:22,23

Only Caleb and Joshua would be permitted to enter the Promised Land.

All others would die in the desert and because of this, "the people mourned greatly." (Numbers 14:39) After 40 years of harsh desert living, the Lord's will came to pass as an entire generation passed. The new generation stood poised to fulfill all that God had promised.

On Giving No Quarter

God's command to Israel, marshaled east of the Jordan River, rings with the decisiveness of divine forbearance. Perhaps it rings of brutality for humanity immersed in the temporal or ignorant of providence concerning the wickedness of the Amorites.

> "When the LORD your God brings you into the land that you are entering to take possession of it, and clears away many nations before you, the (Canaanites)…and when the LORD your God gives them over to you, and you defeat them, then you must devote them to complete destruction. You shall make no covenant with them and show no mercy to them."
> Deuteronomy 7:1-2

Either way, the slaughter of women and children, whether ordained and commanded by God or not, demands consideration and clashes with concepts of mercy and modern just war doctrine. Indeed, protection of non-combatants delineates the savage from the civilized. The valiant warrior protects helpless women and children. It is the savage, the brute, who seeks their harm or places them between him and his enemy. We'll expound on this later but note for now that this is a special consideration, the conquest of the Holy Land, for a special time.

God didn't intend this as a general command concerning warfare, rather a command for this specific instance in history, the invasion of Canaan. He issues other general commands concerning warfare that stipulate mercy and restraint (Deuteronomy 20), just not for this particular invasion.

Prior to Canaan, Israel warmed up with a series of skirmishes against the Amorites. At Edrei, they defeated King Sihon of Heshbon and King Og of Bashan, both of whom impeded Israel's flanking maneuver against Canaan. Israel defeated them soundly and slaughtered everyone—men, women, and children though they spared the livestock and retained it as plunder. (Deuteronomy 2, 3) Forced to circumnavigate Edom, Israel miraculously forded the Jordan River and initiated the invasion from the East. (Joshua 3)

Joshua, as the head of the nation, commenced the culmination of 440 years of anticipated fulfillment of a promise, a covenant with the God of the Universe. God appointed Joshua to the task, selecting him from among the people to succeed Moses. Israel stood ready to consummate their existence. Now was the time.

An Aside Concerning National Israel

Much is made of Israel and rightfully so. God chose them though the Bible quickly points out that God selected them by no merit of their own. They repeatedly turned from God, and still He walked among them. Even while languishing in the wilderness, awaiting the death of an entire generation as a punishment for their rebellion at Kadesh-Barnea, God lived in their presence. National Israel existed and exists today like no other nation.

Consider the unique aspects of Israeli history. They developed while enslaved in Egypt, born as a nation with no actual land. Only upon the fulfillment of God's call did they attain a land for themselves, Canaan. Eventually, they rose to the apogee of their existence under David and Solomon only to see their kingdom torn asunder by the judgment of God. Assyria and Babylon served as the instruments of God's wrath in that regard.

Recall that pagan kingdoms dominated Israel for the remainder of their existence, first Babylon followed by the Persians. Alexander the Great, upon his conquest, infused Greek language and culture into all nations under his dominion, including Israel. Hellenism threatened to dilute Jewish identity beyond recognition. Following a brief time of independence under the Maccabees, Rome subdued Israel and subjugated the Jews for several hundred years. In AD 70, Titus sacked Jerusalem effectively ending a four-year Jewish revolt and terminating Israel's existence as a nation. The few survivors, the remnant, scattered to the ends of the world, joining the Diaspora in exile.

Israel, national Israel, had ceased to exist as a nation and after the Nazi extermination of 6 million Jews in the Holocaust, almost ceased to exist as a people...until May 14, 1948. Following a series of miraculous circumstances, David Ben-Gurion, the head of the Jewish Agency, with United Nations' support, proclaimed the establishment of the State of Israel.

A nation had been reborn in a way unlike any other in the history of the world. Thus, Israel's initial call as a chosen people of God, their dispersion around the globe following the destruction of their nation, the Holocaust, and the rebirth of their nation all remind us of the special nature of Israel.

However, the implications of this special nature are both divisive and difficult to establish conclusively. As they are extraneous to this work, we'll make our previous assertions and move forward.

Consummation of the Conquest

Joshua knew none of this as he marched forward with the full assurance of the righteousness and difficulty of the impending battle. Jericho fell without a hitch as Israel relied fully upon God and his directives. God led, Israel followed, and the first domino in the conquest fell exactly as God ordained. Joshua put Jericho to the sword. He burned it to the ground and exalted in the work of the Lord and as such, "his (Joshua's) fame was in all the land." (Joshua 6:27)

However, things began to fall apart immediately as Joshua moved on Ai, the next city that must fall. Achan, the son of Carmi, took plunder for himself at Jericho in spite of God's warnings not to do so. God's anger blazed against Israel. Consider that God delivered Jericho to Israel, exactly as He said He would. Immediately following the very first victory, the people strayed and so, "the anger of the LORD burned against the people of Israel." (Joshua 7:1b)

Joshua's initial foray against Ai met with failure as he dispatched a much smaller force and they were subsequently routed. "Then Joshua tore his clothes and fell to the earth on his face before the ark of the LORD until the evening, he and the elders of Israel." (Joshua 7:6) However, God revealed their sin and once they addressed Achan's transgression, Joshua sacked Ai, employing a deception to render the city vulnerable.

Ai fell and Israel slaughtered all 12,000 residents. (Joshua 8:25) Joshua burned Ai to the ground and in celebration and remembrance, he built an altar on Mount Ebal and renewed the covenant, reading aloud all that was written in the Book of the Law to the People. "There was not a word of all that Moses commanded that Joshua did not read before all the assembly of Israel." (Joshua 8:35)

Israel remembered God and their campaign continued unabated. Joshua chapters 10 and 11 record the additional conquests of southern and northern Canaan. At Makkedah and Libnah, Lachish and Gezer, Eglon and Hebron and Debir—Joshua struck them with the edge of his sword and the Lord delivered them over to destruction. Joshua faithfully upheld God's commands. He showed no mercy, devoting the Canaanites to complete destruction.

As Canaan progressively fell, the remaining kingdoms increasingly cooperated and consolidated their forces against Israel. However, even consolidation would not save Canaan. At Merom, Joshua and his men fell upon the numerically superior forces of a Canaanite coalition "and the Lord gave them into the hand of Israel." (Joshua 11:8)

Despite these initial successes, this was not to be a quick campaign as "Joshua made war a long time with all those kings." (Joshua 11:18) Canaanite resistance never waned as "there was not a city that made peace with the people of Israel." (v.19)

At some point, the campaign flagged. Israel lost operational momentum and surrendered the initiative. The Bible does not clearly delineate when this happened. After describing the conquest to date, Scripture next records that "Joshua was old and advanced in years." (Joshua 13:1) Joshua became too old to lead his army and much conquest remained. Prior to his death, Joshua set about dividing the land. Israel became content to occupy what they had thus far obtained. (Joshua 24:28)

This contentment and satisfaction with less than what God had propositioned affected the future of Israel irreversibly.

As it Happened

As a direct result of the failed conquest, Israel suffered. The very next verse after our main passage speaks to this failure.

And the people of Israel did what was evil in the sight of the LORD. They forgot the LORD their God and served the Baals and the Asheroth.
Judges 3:7

They forgot within one generation. They forgot God and sought after the false and pagan gods of the Canaanites, just as God had said they would should they not complete the conquest as He directed. Within one generation, the people abandoned God and for that, "the anger of the LORD was kindled against Israel." (v.8)

How may we reconcile Israel's failure? As we have ascertained the source of war and developed the road to war in this particular account, we must now *apply* our theology from Chapter Three. We must overlay the *sovereignty* of God upon the events at hand and develop our understanding of *providence*. Let us examine the apparent tragedy of Israel's betrayal in terms of the conclusion that God *tests* through war as we return to our passage in Judges Three.

6. War and Sovereignty

A First Stab at Concurrence

Returning to the first recorded military campaign, an interesting encounter occurs upon Abraham's return from rescuing Lot. The king of Sodom goes out to meet him as does another figure, Melchizedek, the king of Salem. The Bible calls him, "a priest of God Most High." (Genesis 14:18b) Melchizedek produces bread and wine and blesses Abraham saying,

Blessed be Abraham by God Most High,
Possessor of heaven and earth;
and blessed be God Most High,
who has delivered your enemies into
your hand.
Genesis 14:19-20

In response, Abraham gives a tenth of all that he plundered to Melchizedek.

Melchizedek exists as a mysterious figure in the Bible. Two words comprise his name, Malkîy-Tsedeq, that literally translate as *king of right(eousness)*. The author of Hebrews describes Melchizedek as a "king of peace" (Hebrews 7:2b) and states that "He is without father or mother or genealogy, having neither beginning of days nor end of life, but resembling the Son of God he continues a priest forever." (v.7:3)

Psalm 110, a messianic psalm often quoted in the New Testament, speaks directly to the future Messiah and declares, "You are a priest forever after the order of Melchizedek." (v.4) The author of Hebrews, in lashing Jesus with the Messiah, alludes to Psalm 110:4 in a lengthy dissertation on the priesthood of Christ, that He is "a priest forever after the order of Melchizedek." (Hebrews 7:17)

Christophanies litter the Old Testament, pre-incarnate appearances of Jesus. Clearly, Melchizedek is the Christ. Notice the description in Hebrews—without mother, without father, no beginning, no end, but a priest forever. Hebrews' reliance upon Psalm 110 in establishing Jesus as the Messiah further solidifies the relationship.

Melchizedek's encounter with Abraham proves useful. Abraham and his men just returned from a slaughter, a great military victory. Dried blood likely still stained their swords. Having traveled and fought all night, they are exhausted, yet exhilarated at the victory, drained but quietly joyful. Melchizedek does four things:

1. He brings out bread and wine, foreshadowing the ordinance of Communion.
2. He blesses Abraham.
3. He blesses God Most High.
4. He receives a tithe (tenth) offering from Abraham.

Abraham, who just shed blood serving the purposes of men, receives the blessings of Jesus, the blessings of God. He fought fiercely but honorably as, at the command of the LORD, he took no plunder. The text stops short of a definitive causal relationship but associates the blessings from God with the fact that Abraham fought not to serve jealousy or selfish ambition. Rather, he fought at the pleasure of the Lord. Tellingly, Abraham further honors God with a tribute in the form of a tithe.

Notice the concurrence. Abraham decided to execute the raid to free his nephew. Abraham divided his forces and led them during the highly successful assault. Yet, Melchizedek declares to Abraham that God "has delivered your enemies into your hand!" (Genesis 14:20) Abraham freely executed the mission prompting a reward from God for his faithful execution. God dwells in the shadows, behind the curtain, crafting the will of men and events to serve His purposes. Just as God holds men accountable for their wicked actions, God rewards the faithful commitment of men, though He is Himself the primary cause of their actions, in the spirit of concurrence.

"Oh the depth of the riches and wisdom and knowledge of God! How unsearchable are his judgments and how inscrutable his ways!" (Romans 11:33)

On the Concurrence of War

Our paradigm governs our comprehension of the relationship between the sovereignty of God and war. Through which lens do you view the world? We've noted from texts such as Romans 8:28 that all things work together for the good of those who love God and ultimately for His glory. Notice the text refers to *all* things, not some things, not many things, but all things. This must include war.

God enslaves all things to His decree, for the good of His people. Israel's failure to complete the conquest did not surprise God. Israel's failure to completely drive out the Canaanites did not catch the sovereign Lord with His guard down. It's not as if He turned to the Holy Spirit and said, "Well, what do we do now?" Examine verse 1 of our passage in Judges,

> *Now these are the nations that <u>the LORD left</u>, to test Israel by them, that is, all in Israel who had not experienced all the wars in Canaan.*
> *Judges 3:1*

The LORD is who left the nations, the Canaanites. At this point the reality of the paradoxical God confronts our thinking. Joshua left the Canaanites. Israel left the Canaanites. They own the failure and God holds them accountable because He repeatedly warned them of the consequences of an incomplete conquest.

From a macro viewpoint of the salvation history of man, they never anticipated the unlikely outcome. They possessed no prescience. They didn't know God's secret will just as we do not. They failed due to their own lack of faith and fortitude. Israel left the Canaanites in the Promised Land, the very people that God had commanded them to drive out completely lest they corrupt all things; Israel left the nations.

The LORD left the nations. See the beautiful, irreconcilable complexity of God's providence, associating and harmonizing the will of God with the biblical free will of moral agents.

> *So the LORD left those nations, not driving them out quickly, and he did not give them into the hand of Joshua.*
> *Judges 2:23*

This verse from the previous chapter confirms this idea. God authored the invasion. God directed. God ordained. This invasion was planted in the hearts of men by an all-knowing God. God commanded them to fight then, God fought on their behalf.

> *it was not by your sword or by your bow.*
> Joshua 24:12

Consider the implications for a moment if you will. Consider all that Israel had accomplished. At the command of the LORD, Israel sacked Jericho and Ai. Language proves decisive.

The execution of the five Amorite kings belonged to the LORD. (Joshua 10:16-28) The strong arms that rolled the stone in front of the cave, trapping them inside, belonged to the LORD. (v.17) The rough hands that brought them before Joshua, thrusting them to the ground, belonged to the LORD. (v.24) The firm boots that pinned their necks belonged to the LORD. (v.24) The strong hand of Joshua that struck them down and hung them on a tree for all to see belonged to the LORD. (v.26) Envision Joshua's bloody sword, stamped on one side, "property of Joshua son of Nun" and on the other, "property of the LORD our God."

Brave Warrior, in much the same way, the trigger you pull belongs to the Lord. The trigger pull itself is a slave to the will of the Lord, Almighty. God enslaves the 7.62 as readily and as easily as he enslaves the 40mm, as readily and easily as He enslaved the Assyrian nation. God enslaves the American as assuredly as He enslaves the jihadi or the German. All things and people serve His will and purposes, including all things that comprise this harshest of relational interactions, the thing that is war.

Knowing this, how will you pull the trigger? In what way and with what assurance do you adjust your sight picture? Far from forsaking or abandoning all things with flippancy, "Well, it'll serve God anyway so I'll conduct myself as I please." No, shift to a different trajectory. Because this thing that I do ultimately serves the Lord, because this thing that I do ultimately serves Him and His purposes, I'll conduct myself with that much more gravity. Never will I pull the trigger with disregard. Never will I pull the trigger with misunderstanding or brazenness.

No, I will pull the trigger in the full confidence of the sovereignty of the LORD our God, seeking to enslave my purposes to His will all while knowing that I have no actual choice in the matter. My purposes already serve His will. For that reason, I will seek the same, that which already exists. I will draw my sights as from the soul, as something done unto the Lord and not for men.

God's sovereignty becomes even clearer as we examine further the idea of God testing through war.

Of Obedience

"Will you fight?" God inquires.

We see in this idea of God testing, a confrontational test for obedience. "Will you fight?" God asks yet again. NOT fighting is frequently a much more palatable course of action.

It was this spirit, the spirit of appeasement and moral ambiguity that prompted Chamberlain to declare, "Peace for our time," after ceding Czechoslovakia to Hitler in the hopes of averting another European war.[25] This spirit prompts any concession to evil for the sake of peace. With history as our instructor, we see that evil can never be appeased. Knowing this, will we yet fight?

Fighting Canaan proved difficult for Israel, not the easier thing. Fighting is hard. Soldiers get wounded. Soldiers die. War always yields trauma, bloodshed, affliction, and cruelty. War is nasty business, perhaps the nastiest of businesses and in many cases, it's just easier not to fight. It's easier to seek peace no matter the cost. Israel chose the easier path in conceding to Canaan instead of pursuing their destruction as God had ordered. Just leave well enough alone.

I can hear their hearts, "We conquered most of the Holy Land, plenty of land for everyone. The rest of these folks, including the Gibeonites, should make good neighbors and look at their daughters. They have beautiful daughters and we'll take them as wives and give them our daughters as wives and we'll still worship our God, we'll just include theirs as well."

Painless assimilation beckoned, tempting their concession. Knowing this, prior to the invasion, God commands Joshua and Israel,

Be strong and courageous,...
Joshua 1:6

and one verse later,

Only be strong and very courageous,...
Joshua 1:7

As war is a nasty business, it requires strength and courage. Physical strength proves useful in war but here, God speaks to another strength,

---

[25]"Neville Chamberlain on Appeasement (1939)," *the History Guide*, accessed March 24, 2016, http://www.historyguide.org/europe/munich.html.

- *strength of character,*
- *strength of faith,*
- *strength of resolve.*

No one ever said, "be strong and courageous in order to surrender to the enemy or appease this great evil." It is the waging of war for the sake of victory that requires strength and courage. Few things test a man's mettle more than war.

In Deuteronomy 20, God provides general laws concerning warfare before issuing some specific commands concerning the conquest of the Holy Land. The text envisions the tough fight to come, "when you go out to war…and see horses and chariots and an army larger than your own." (Deuteronomy 20:1) When this happens, God exhorts them, "you shall not be afraid of them, for the LORD your God is with you." (v. 1b)

Prior to battle, God requires that the priests speak to the people, exhorting strength and courage. He requires that all warriors be fully vested, completely committed. If any possess a distractor, either a house that he needs to dedicate or a vineyard he has yet to harvest or even a new wife he has not 'taken', then that man should return and take care of the necessary business. God seeks the full engagement of the individual warrior, complete resolve.

Next, God directs the officers to speak to the people saying "Is there any man who is fearful and fainthearted? Let him go back to his house, lest he make the heart of his fellows melt like his own." (Deuteronomy 20:8) Cowardice is contagious. God demands repeatedly, be strong and courageous even to the point of removing from the ranks those who might falter, those who possess a weak heart. The conduct of war demands resolve of the men who wage it.

Lastly, as the commanders address the people, God prescribes the terms of the offensive. He requires overtures of peace prior to the assault and describes the terms of capitulation, delineating actions should the besieged people refuse the terms. He even directs the terms of the termination of hostilities and post-combat actions. (vs.10-15)

About Fear

Not that I'm any sort of steely-eyed killer, but I've only actually been afraid in battle a handful of times. Normally, the speed and efficiency of our actions precluded fear. There's simply no time for fear. Usually, it's in the form of post-mission reviews that I conclude, "Wow, that was not the smartest thing we've done. We probably shouldn't be alive right now." Retroactive fear if you will.

We just landed a large helicopter in the middle of the road in broad daylight in bad guy land. We shouldn't be alive right now.

We just hovered a helicopter over a car full of bad guys with RPG's. We shouldn't be alive right now.

We just executed a vehicle takedown at high noon in downtown Mosul. We shouldn't be alive right now.

Live we did, either by the grace of God or the ineptitude of the enemy or in the spirit of concurrence, both. These unplanned actions, had I the opportunity to consider in advance, would have generated requisite angst, and rightfully so. God nowhere says, "Be an idiot and deliver yourself into the hands of your enemy", at least that I can find.

War tested my mettle and my resolve. I recall my first tour in command of a task force in Iraq and all the angst in my heart during the 20-hour trip to get there. I recall a pre-planned hasty daytime assault into the Panjwai River valley just west of Kandahar—a virtual hornet's nest of bad guys—and the lump in my throat, adrenaline surging my veins as we called 'up' and launched to the objective. I recall a similar planned assault into the Helmand River valley, attacking a compound of bad guys south of Marjeh and my intense discussion with the ground force commander concerning the utility of such a venture. Is this even a good idea? Does the benefit outweigh the cost? Is the juice worth the squeeze? Could I look into the eyes of one of my soldier's wife at his funeral and tell her it was worth it?

Though I never personally witnessed a single soldier falter, fear of personal harm underscored even our mission planning, though group dynamics and norms greatly inhibited public manifestations. At some point, I developed a decidedly fatalistic, though not entirely unbiblical attitude that the Lord will take me when it's my time and not a second sooner. A much greater fear gripped my heart, that I would let my buddies down or my soldiers. Would I crumble during the intensities of combat or would I stand?

Thus, the Lord's exhortation to have courage and strength. Few things tax the soul like the fatigue of battle—a physical, spiritual, and emotional fatigue that sinks to the bones. The friction of battle makes things, even seemingly simple and ordinary things, indescribably difficult. We fight to win and we slay ourselves on the ever-turning wheel of mission at the expense of combat power always balancing preservation of combat power with the accomplishment of said mission, a tenuous and strained relationship that possesses no actual resolution.

In northern Iraq once, we finished a long night of operations and were flying back to our operating base for some hot chow and some much-needed sleep when we got the call. Another ground force had located some bad guys of renown and were ready to engage, time now. After a brief discussion, we diverted our flight to another base, picked up a ground force, and launched toward the objective area. We planned the mission during the one hour flight and oh yeah, another helicopter assault force would meet us on the assault as the objective was deemed large enough to require both forces.

After an already 20-hour day, we had to coordinate and execute a complex, daytime helicopter assault against a fortified position with two geographically dispersed forces, on the go, over the radio. Somehow, we executed and didn't crash any helicopters into each other. The ground force assaulted the objective, we extracted them and recovered to a temporary staging base. I don't even remember if we got the bad guys or not. After more than 30 hours of combat, I didn't remember much.

What I do remember is the sleep tent. I rarely sleep better than in an Army cot, and some loggy had procured a sleep tent full of cots. It was pitch black inside and against the bright heat of northern Iraq, ice cold. I slept like the dead for about 18 straight hours. We ended up executing another assault on the way back to our home base the very next evening. God tests through war and,

1. Fighting is hard. It's often easier not to fight,

2. God commands men to be strong and courageous.

As God tested Israel, we conclude that fighting evil is hard and God tests our obedience, our willingness to partake in this most difficult activity. He tests our courage and strength, our mettle. Will we meekly fold at the advance of evil or will we stand with the full authority of the God of the Universe and declare, "Not on my watch"?

The question of whether a Christian ought to fight sort of fades into irrelevancy in view of God's sovereignty and his call to confront evil collectively. I would argue that the Christian, because he fights on behalf of a higher purpose in consort with the Lord, fights that much harder, with that much more tenacity and determination. Though I may fall on the battlefield, I'll never fall in vain. Rather than fall, the Lord will have laid me gently to rest as I'll awaken to eternity and the confidence of hearing, "well done, good and faithful servant." (Matthew 25:23)

Disobedience

Israel was tested and found wanting. As our text indicates, "they (Canaan) were for the testing of Israel, to know whether Israel would obey the commandments of the LORD, which he commanded their fathers by the hand of Moses." (Judges 3:4) God issued these difficult commands in part, to see if Israel would obey, to test them. They failed, and history yields extensive consequences for that failure. As it *should have been written*,

> Israel, emboldened by the LORD, completely vanquished the Canaanites and took possession of the entirety of the Holy Land just as God had commanded and lived out their days in worship, fellowship, and service to the LORD.

As it actually happened,

> *So the people of Israel lived among the Canaanites, the Hittites, the Amorites, the Perizzites, the Hivites, and the Jebusites. And their daughters they took to themselves for wives, and their own daughters they gave to their sons, and they served their gods.*
> *Judges 3:5, 6*

Surely many objected to the harshness of God's commands concerning the invasion.

> *you shall save nothing alive nothing that breathes, but you shall devote them (Canaan) to complete destruction, ...*
> *Deuteronomy 20:16b, 17*

This seems excessively harsh, unloving, and unjust. The Bible, however, speaks extensively to God's justice. In His perfection, all of His decrees display perfect justice.

> *For I the LORD love justice,*
> *Isaiah 61:8*

> *Of a truth, God will not do wickedly,*
> *and the Almighty will not pervert justice.*
> *Job 34:12*

> *...for all his ways are justice.*
> *Deuteronomy 32:4*

Perhaps we have trouble understanding. Perhaps we have trouble reconciling, but all of God's actions provide perfect justice. He simply cannot act in an unjust manner as that would betray His very character. Now, in salvation, God elects some to receive mercy and as such, all men receive either justice or mercy. For those granted mercy, Jesus bears the brunt of God's justice on their behalf.

There is perfect justice in His commands concerning Canaan. Consider the rationale He provides,

> *you shall devote them to complete destruction....that they may not teach you to do according to all their abominable practices that they have done for their gods, and so you sin against the LORD your God.*
> *Deuteronomy 20:17, 18*

God's purposes transcend the temporal. Knowing the fickle hearts of his people, God seeks their purity and holiness, desiring to eliminate all that would corrupt. Previously we saw that in mercy, God delayed the invasion for 400 years to accommodate the Amorites (Canaanites) that they might repent and turn to the Lord. God enslaved His people in Egypt for 400 years, in patience, on behalf of Canaan. In mercy, God suffered His people for those who denied Him and would likewise tempt His people to deny Him.

Now, though, the iniquity of the Amorites was finally full and their destruction imminent. In Israel's failure notice that it happened exactly as God said that it would, in their disobedience.

The allure and seductiveness of evil, of the world, leads countless to appease, to tolerate, to assimilate. The Canaanites did not yield to Israel, fighting valiantly and so Israel disobeyed God and settled with their enemy, choosing peace. Peace is good! Right? Well, perhaps peace when it means disobedience to God and compromise with evil is not so desired.

Notice what the text says, that they (Israel) lived among the Canaanites. Israel, rightful owners of the land as promised to Abraham, lived among them, the Canaanites, indicating the extensive nature of their failure. There were enough Canaanites left to state that Israel lived among them.

Even further, our text states that "their daughters they took to themselves for wives, and their own daughters they gave to their sons, and they *served their gods.*" (v.6) Israel not only lived amongst the Canaanites, but they intermarried with the Canaanites and inevitably abandoned the God of salvation and served Canaanite gods, pagan gods—Molech, Baal, and Asherah.

But peace is good, right? In this case, peace at any cost led to judgment as the rest of the Old Testament drives this narrative to its completion, playing out the drama to which we already know the end. Israel could never contain the spiral into sin that began with their disobedience. Eventually, they fell into syncretism, the blending of the worship of the one true God with the worship of false gods. They introduced temple prostitutes into the very House of God in Jerusalem. They sacrificed their children in fire to the false god Molech. The iniquity of the Amorites became Israel's iniquity.

In *mercy*, God sent prophets, repeatedly. Isaiah, Jeremiah, and Ezekiel all came to call God's people to repentance. Several times the people relented and a time of national revival ensued, but always they returned to their sin. Hosea, Joel, Amos, Obadiah, Jonah, Micah, Nahum, Habakkuk, Zephaniah, Haggai, Zechariah, and Malachi—God sent them all to call His people to account. Mostly their entreaties fell upon deaf ears.

In mercy, God offered time, delaying His judgment for centuries until in 722 B.C., He raised up Assyria to destroy Israel and in 586 B.C, He deployed Babylon to destroy Judah. Ever faithful, God retained a remnant of the nation and eventually returned them to Canaan, but they existed in subjugation for the rest of their existence until the great bloodletting of A.D. 70.

All of this stemmed from disobedience. God called them to fight. They chose peace, peace with evil. Unfortunately, nothing evil desires peace. Evil knows no peace. God knew the hearts of His people and He knew that a total conquest of Canaan was necessary, lest they fall.

Conceptually, many shy away from objectivity. As we've noted, most people desire subjectivity. The Bible offers nothing less than objective truth. Evil exists. There is evil in this world—evil forces, evil spirits, wicked men governed by the evil resonate in their hearts. We have an enemy and the enemy cannot be appeased. The enemy cannot be negotiated with. The enemy may take different forms, look differently at different times throughout history, but always the enemy seeks after nothing less than total destruction, physically and spiritually.

Consider the current enemy of the west, Islam. The very name of Islam means submission. Islam seeks nothing less than to subjugate the entire planet via force if necessary, imposing sharia law and eliminating *all* competing ideologies, especially Christianity. Though the bulk of lukewarm Muslims seem to offer the notion of a peaceful coexistence, the devout Muslims (thankfully in the minority) soundly refute this.

I pray that our nation would not be tested and found wanting, that brave warriors, men of strength and courage, would continuously run to the proverbial sound of the guns and fight, never relenting to evil and the wickedness of men. As God tests through warfare, the consequences for our future disobedience are as obvious as they are imminent.

War Defined

Behind the testing of Israel, God's sovereign hand guides, directs, and ordains in providence. Concurrence, in view of the Fall which generated the conditions that now govern existence, drives us to augment our definition of war.

> *War is the systematic corporate friction between the collective wills of sinful men otherwise enslaved to the will of God.*

The necessary addition encompasses enslavement. Enslavement prompts a reconciliation of the revealed will of God with His secret will which truly governs. As war exists, it must necessarily serve His purposes. God enslaves war as all things exist in servitude to His authority.

Though war derives from the incorporated impure motives of men, we may not declare it as illegitimate. The concept of God testing and teaching through war assures us of this. Let us turn to Christ to bolster this conclusion.

Chapter 7

7. Lambs and Lions

I have a well-studied friend who likes to ask me difficult questions concerning the Bible. A stoutly built redhead normally adorned with a menacing scowl, he at some point picked up the Bible and started reading because, as he put it, "I want to figure out how it all started." I assumed he was speaking about our current conflict.

Despite my exhortation to read the Bible in its entirety, he remained fixated on the Old Testament and as he consumed the text he would drill into the most peculiar details and then seek a consultation. I always felt ill-prepared to answer. We were spending quite a bit of time just north of Baghdad and he'd approach me, usually in the operations center, with a snarl and gruffly inquire,

"Brad, when Moses anointed Aaron and his sons, why did he rub blood on the lobe of his right ear, his right thumb, and his right big toe."

"Brad, what's up with the song of Moses in Deuteronomy? Pretty intense stuff. What does it all mean?"

"Brad, why don't we obey the Old Testament dietary laws?"

"Brad, why is the God of the New Testament different than the God of the Old Testament." He had finally taken my advice and cracked the New Testament and happened upon a seemingly glaring inconsistency, one that skeptics have leveraged since the New Testament was written.

The God of the Old Testament is a God of wrath and judgment, different from the God of the New Testament who is a God of love, kindness, and mercy. Interestingly, it is the New Testament God of love that most cling to forsaking the Old Testament God of wrath as a God of antiquity, a God reserved for a bygone epoch.

Most couple this view with our previously mentioned effeminate Jesus. Let us assess "both" Gods for consistency and culminate with a comparison of the affable Jesus with the tyrannical God of the Old Testament.

Congruency between the second person of the Trinity and the frequently affirmed harshest of the three, the Old Testament God, would seem to eliminate any substantial reason for skepticism on the basis of inconsistency.

The Divine Paradox

Merriam-Webster defines a paradox as, "something (such as a situation) that is made up of two opposite things and that seems impossible but is actually true or possible." I'd like to visit the divine paradox that is the Father as described in Exodus 34.

> *The LORD, the LORD, a God <u>merciful</u> and <u>gracious</u>, <u>slow to anger</u>, and <u>abounding in steadfast love</u> for thousands, <u>forgiving</u> iniquity and transgression and sin, but who will <u>by no means clear the guilty</u>, <u>visiting the iniquity</u> of the fathers on the children and the children's children, to the third and fourth generation. Exodus 34:6,7*

We see, in this description of the LORD, that He is merciful. He is gracious. He is slow to anger. He is abounding in steadfast love. He forgives iniquity and transgression and sin. This reconciles well with popular connotations of the God of the New Testament, of Jesus. People want to believe this about God. People always believe this about God.

You've heard it said, "God loves me. God loves all His children." The paradigm today is much different than a century ago. In that time, a preacher's challenge resided in convincing people that God could love them in spite of their sin. Today, you tell someone that God loves them and they will likely respond, "Of course God loves me. He loves everyone and I'm a good guy." People have completely embraced the God of love, the God of mercy, the God of graciousness.

Likewise most have rejected the latter half of this passage, the aspect of wrath. Consider what this passage actually says. God will <u>by no means</u> clear the guilty, meaning never, under any circumstances will He allow the guilty to walk free, unpunished. He will visit the iniquity, the punishment for men's guilt, unto their children and their children's children. Punishment for iniquity will span generations.

Well, who are these guilty people? The Bible answers this easily enough.

> *For all have sinned and fall short of the glory of God, Romans 3:23*

None is righteous, no, not one;
no one understands;
no one seeks for God.
All have turned aside; together they
have become worthless;
no one does good,
not even one.
Romans 3:10-12

Paul writes clearly that all have sinned, no one is righteous. All men find themselves not only accused but guilty before the Lord. All men, unfortunately, find themselves in the latter half of the paradox. We all stand guilty before God, and God will <u>by no means</u> clear us of this guilt! He will punish sin. He must punish sin. His very nature demands that He punish sin. Thus, Isaiah writes,

I will punish the world for its evil, and the wicked for their
iniquity;
Isaiah 13:11

For behold, the LORD is coming out from his place
to punish the inhabitants of the earth for their iniquity, and the
earth will disclose the blood shed on it, and will no more cover
its slain.
Isaiah 26:21

For by fire will the LORD enter into judgment, and by his sword,
with all flesh; and those slain by the LORD shall by many.
Isaiah 66:16

The reality of what the Bible actually says startles in its implications. Is God merciful or is God just? The fullest extent of this line of reasoning yields the query, "What about me? What happens to me? What will God do with me?" A nod to the previous passages prompts a telling response,

"That is so Old Testament. That is not what *my* God is like. My God would never hate. He is love." Many earnest Christ followers cling to this idea.

Concerning the Wrath of God

God hates the sin but loves the sinner, a mostly insipid and fallacious claim. The actual truth challenges a bit more. God hates sin and in many respects, sinners with equal aplomb. Absorb this truth into the very essence of your being. Recall *who* it is that hates. The all-powerful, all-present, all-knowing God of the universe hates sin and even sinners, and to its fullest extents, hates the very fiber of your essence.

> *The boastful shall not stand before*
> *your eyes; you hate all evildoers.*
> *Psalm 5:5*

> *The LORD tests the righteous, but his soul hates the wicked and*
> *the one who loves violence.*
> *Psalm 11:5*

> *There are six things that the LORD hates, Seven that are an*
> *abomination to him: haughty eyes, a lying tongue, and hands*
> *that shed innocent blood, a heart that devises wicked plans, feet*
> *that make haste to run to evil, a false witness who breathes out*
> *lies, and one who sows discord among brothers.*
> *Proverbs 6:16-19*

Consider that as Israel marshaled their forces at Gilgal after crossing the Jordan River, even there, God began to hate them.

> *Every evil of theirs is in Gilgal; there I began to hate them.*
> *Hosea 9:15*

He began to hate His people for the sin and disobedience already resident in their hearts, sin that had not yet become action! Speaking of the Canaanites,

> *And you shall not walk in the customs of the nation that I am*
> *driving out before you, for they did all these things, and*
> *therefore I detested them.*
> *Leviticus 20:23*

As sin polluted and corrupted your very soul, your very being, God hates you. You have turned from Him, you have rejected Him, and you have hated Him. You are a depraved wretch and God loathes and detests your very nature. This is a very unsettling thought.

God's holiness demands that He hates sin and it is impossible to disassociate the sin from the sinner. God cannot punish sin. Consider a man brought before a judge for the commission of some offense. Could the judge punish the offense but not the offender? No, it is the offender who bears the brunt of justice for the commission of the offense.

It is the same with sinners standing before a holy and righteous God and He is an angry God.

Even at Horeb you provoked the LORD to wrath, and the LORD was so angry with you that he was ready to destroy you.
Deuteronomy 9:8

Now, therefore, let me alone, that my wrath may burn hot against them,
Exodus 32:10a

And the LORD's anger was kindled against Israel,
Numbers 32:13a

You marched through the earth in fury; you threshed the nations in anger.
Habakkuk 3:12

As God is angry, consumed with wrath for the wicked, He is jealous, demanding complete loyalty and obedience.

for you shall worship no other god, for the LORD, whose name is Jealous, is a jealous God,
Exodus 34:14

...He is a jealous God; he will not forgive your transgressions or your sins.
Joshua 24:19b

How long, O LORD? Will you be angry forever?
Will your jealousy burn like fire?
Psalm 79:5

His first two commandments speak to exclusiveness, even demand it. "You shall have no other gods before me," (Exodus 20:3) and you shall not worship an image, a created thing, the essence of idolatry. (vs.4-6) You shall not even misuse the name of the LORD. (v.7) God is no benevolent bystander, content to allow men to trample and profane His holiness, mocking His position as LORD of all.

> *The LORD is a man of war;*
> *the LORD is his name.*
> *Exodus 15:3*

God is a fierce warrior, hating all that is unholy and unrighteous and He will punish the wicked. He will vanquish evil. He will visit wrath for the sins of men upon their very souls.

God's actions repeatedly affirm this. Consider that as evil spread across the land, he destroyed the entire earth with a flood. He destroyed men, women, and children, drowning them in their iniquity, by the thousands, the tens of thousands, even the hundreds of thousands. Dare to consider taking a baby and holding it underwater until it dies, seeing the terror in its eyes and the astonishment as it goes to draw breath but instead finds its lungs full of water, kicking in futility at your arms with no hope of escape.

I cannot consider such a notion. Now consider, for those of you who have children, maybe even a baby at present, doing that very thing to your own. Dare to consider that God did that very thing on a macro level in His anger, as His fierce wrath burned against the sin of men.

On a smaller scale, consider Nadab and Abihu, sons of Aaron, who "offered unauthorized fire before the LORD, which he had not commanded them." (Lev. 10:1) Scripture does not detail what sort of unauthorized fire but the LORD had not prescribed it. Did He rebuke them? No, He killed them. "And fire came out from before the LORD and consumed them, and they died before the LORD." (v. 2) So serious is God about obedience and holiness that He instantly destroyed two men, even as they sought to serve Him.

Consider Uzzah. As Israel defeated the Philistines and reclaimed the ark, they brought it home in a great procession to wild celebration. As they carried the ark on a cart drawn by oxen, at some point the oxen stumbled and Uzzah reached out to steady the ark and keep it from falling. "And the anger of the LORD was kindled against Uzzah," (2 Samuel 6:7a) and did He rebuke him? No, He killed him. "God struck him down there because of his error, and he died there beside the ark of the God." (2 Samuel 6:7b) Uzzah disregarded the direct command of God, assuming that his hands were less filthy than the ground, and paid for it with his life.

Consider that in 722 B.C. God raised up the nation of Assyria and deployed them against His people as they destroyed the northern kingdom of Israel. Consider that in 586 B.C. God likewise raised up the Babylonians and deployed them against His people, the southern kingdom of Judah. Consider that God used both Assyria and Babylon as instruments of His divine wrath, the rod of His fierce anger as He trampled the winepress of His fury and judgment.

"It is a fearful thing to fall into the hands of the living God." (Hebrews 10:31) Perhaps no one has uttered a truer statement than this.

Concerning the Love of God

We need not establish, with too much rigor, the wrath of the 'Old Testament' God. Everyone already acknowledges this with most assuming away this aspect of His character.

In spite of the sheer volume of Old Testament passages concerning the wrath of God, we find oppositional accounts. He judges with reluctance always and He judges with purpose. As He declares to Ezekiel, "I have no pleasure in the death of the wicked, but that the wicked turn from his way and live." (Ez. 33:11) Even in judgment, God seeks after the hearts of men and as such, we will not neglect an examination of the first half of the paradox, His mercy, grace, and love.

Look no further than the original promise to Abraham, the father of His chosen people, precursor to the Church. God tells Abraham, "I will make of you a great nation, and I will bless you and make your name great," (Genesis 12:3) and for what purpose does He tell him this? "So that you will be a blessing," (v.2b) He tells Abraham, "in you all the families of the earth shall be blessed." (v.3b) God chose to bless Abraham so that this blessing would extend to all families on the earth.

We know, by viewing the Old Testament through the lens of the New that God spoke of Christ, that Christ would be the blessing to all the families of the earth. We see exactly this today. Even those who deny Christ enjoy the blessings of common grace.

God's passion and love for His people stands in stark contrast to His wrath and anger. In light of impending judgment, God declares,

How can I give you up, O Ephraim?
How can I hand you over, O Israel?
My heart recoils within me;
my compassion grows warm and tender.
Hosea 11:8

God has a passion for His creation, for people in general, and a special love and compassion for His people evident in His anguish at judgment. As stark as the language of the Old Testament is concerning the wrath of God, it often speaks in intensely passionate language concerning the love of God.

> *Is Ephraim my dear son?*
> *Is he my darling child?*
> *For as often as I speak against him,*
> *I do remember him still.*
> *Therefore my heart yearns for him;*
> *I will surely have mercy on him.*
> *Jeremiah 31:20*

Even in judgment God remembers and loves His people. Israel is like a 'dear son' to Him, a 'darling child'.

God loves the downtrodden, the oppressed, the burdened, and the poor.

> *The LORD sets the prisoners free;*
> *the LORD opens the eyes of the blind.*
> *The LORD lifts up those who are bowed down;*
> *the LORD loves the righteous.*
> *The LORD watches over the sojourners;*
> *he upholds the widow and the fatherless,*
> *Psalm 146:7b-9a*

God champions the dispossessed and commands His people to join Him in loving them and caring for them. The New Covenant displays the greatest expression of God's love in the Old Testament in looking ahead to Christ.

> *I will make with them an everlasting covenant, that I will not*
> *turn away from doing good to them. And I will put the fear of me*
> *in their hearts, that they may not turn from me.*
> *Jeremiah 32:40*

He will not turn away from doing good to them and this is an everlasting covenant, an everlasting condition. Despite the affliction of their sin, God provides hope of something better. God is angry at His people *because* of His love for them. Had He no regard for men, He'd have no regard for their rebellion. Had He no regard for men, He'd never love them enough to send His Son unto them.

Concerning the Love of God (cont.)

Without too much trouble, we transition to the New Testament God of love. Again, most easily embrace this God. It comforts us as,

...God is love.
1 John 4:8

God is love. He is the very essence of love and because He is love, He commands His people to love. The most recognized Bible verse speaks to the love of God,

For God so loved the world, that he gave his only Son, that
whoever believes in him should not perish but have eternal live.
John 3:16

How much does God love the sinner? He loves the sinner so much that it pleased Him to crush His very Son and to pour out His wrath for the sins of the world upon Christ at the cross. The cross is the nexus of the wrath of God and the love of God.

To see the wrath of God, look to the cross as God pours out his hatred and loathing and fierce anger for the sins and rebellion of men upon the Son, crushing Him, piercing Him....all because He loves men, loves sinners.

If you want to see the love of God, look to the cross as Jesus laid down His life for sinners, bearing the justice of the world upon his broken and frail body. As Paul writes,

But God shows his love for us in that while we were still sinners,
Christ died for us.
Romans 5:8

Meditate upon this truth, this love. I beseech you that you have rejected God in your heart, at some point in your life, even now. It is you who deserves the cross. Christ went willingly. Did He go for you? Did He perish for you? Is your name included among the reconciled or those yet to be reconciled?

because, if you confess with your mouth that Jesus is Lord and
believe in your heart that God raised him from the dead, you will
be saved.
Romans 10:9

The Bible maintains that all who believe will be saved. This is the great love of God, that Jesus paid the penalty for sin, once and for all. As we transition to a broader understanding of salvation and the person of Christ, we quickly encounter more uncomfortable truth.

Of the Wrath of Christ

Exclusion provides the first indicator that all is not as we'd wish. Perhaps few things offend more than exclusiveness. Countless people exclaim, "There can't be just one way!" Though numerous biblical writers address the truth, the words of our Savior speak most directly.

> *Jesus said to him, 'I am the way, and the truth, and the life. No one comes to the Father except through me.'*
> *John 14:6*

Who can come to Father apart from Jesus? Who can come to God apart from Jesus? Well, according to Jesus, no one. No one can come to God. No one seeks after God. No one pursues God. Apart from Christ and belief in His atoning work on the cross, all are condemned. This uncomfortable reality confronts our very sensibilities.

What about the other ways?
What about other religions?
What about me? I'm a good person.

The words of Jesus stand true for eternity against these considerations. This is the New Testament God of love, right? Surely the 'real' God of the New Testament would forgive everyone. Right? Paul again confronts this,

> *Let no one deceive you with empty words, for because of these things the wrath of God comes upon the sons of disobedience.*
> *Ephesians 5:6*

The 'empty words' that Paul speaks of are that God will not punish evil doers, the wicked and as Paul previously declared, "ALL (emphasis mine) have sinned and fall short of the glory of God." (Romans 3:23) Further,

> *For the wrath of God is revealed from heaven against all ungodliness and unrighteousness of men, who by their unrighteousness suppress the truth.*
> *Romans 1:18*

That's God, though? What about Jesus? Jesus is friendly and nice. "Jesus is my friend," is the contemporary proclamation. As we've already established in our presuppositions the deity of Christ, we need not examine further since all passages that apply to God the Father also apply to God the Son unless an ontological reason demands separate consideration. However, for the sake of argument, we'll entertain this notion concerning Jesus.

From the first to the last, the Bible speaks in very vivid terms concerning the work and character of Jesus. Genesis 3:15 records the first preaching of the Gospel, interestingly preached to Satan. God says to Satan,

> *I will put enmity between you and the woman,*
> *and between your offspring and her offspring;*
> *he shall bruise your head, and you shall bruise his heel.*
> *Genesis 3:15*

God declares that Jesus, the offspring of the woman, will one day bruise (crush) the head of Satan. His work, His labor, is one of violence, judgment against evil.

To Abraham, God promises to bless all the nations, speaking of Jesus. God will bless all nations by sending Jesus to the world through the line of Abraham. Does He bless them by sending Jesus as our friend?

Jacob blesses his sons from his deathbed. Of Judah, from who runs the line of Jesus, he says,

> *Judah, your brothers shall praise you;*
> *your hand shall be on the neck of your enemies;...*
> *The scepter shall not depart from Judah,*
> *nor the ruler's staff from between his feet,*
> *until tribute comes to him;*
> *and to him shall be the obedience of the peoples.*
> *Genesis 49:8,10*

Not exactly the harmless image of Jesus painted by modern day heretics. His hand shall be on the neck of His enemies. Wait, I thought He was my friend? With the scepter, He shall rule until tribute comes to him or as an alternate rendering "until he comes to whom it already belongs." It belongs to Him, all of it, and He has already come to stake His claim, to declare that which already is.

He will return again, to finalize this claim, to consummate it. No matter your eschatological slant, the Bible says that in the last days, Jesus will return to the earth visibly and tangibly. The Apostle John writes about the Second Coming of Christ,

> *Then I saw heaven opened, and behold, a white horse! The one sitting on it is called Faithful and True, and in righteousness he judges and makes war. His eyes are like a flame of fire, and on his head are many diadems, and he has a name written that no one knows but himself. He is clothed in a robe dipped in blood, and the name by which he is called is The Word of God. And the armies of heaven, arrayed in fine linen, white and pure were following him on white horses. From his mouth comes a sharp sword with which to strike down the nations, and he will rule them with a rod of iron. He will tread the winepress of the fury of the wrath of God the Almighty.*
> *On his robe and his thigh he has a name written, King of kings and Lord of lords.*
> *Revelation 19:11-16*

Jesus will return one day in power and fury, and John says that "He judges and makes war!" He judges and makes war! Our kind and soft Jesus, our mellow Jesus, our friend, He will return to judge and to make war. John describes Him in all of His glory, stunning and righteously angry and He will strike down the nations, putting all of His enemies under His feet. He will rule with a rod of iron, treading the winepress of His fierce fury. He is the King of all kings, the Lord of all lords!

This is the Jesus of the Bible. This is the same Jesus who went to the cross. He came as a lamb, as the Lamb of God, but He'll return as the Lion of the Tribe of Judah. This is the Jesus I serve, not some sentimental idea generated in the weak and tepid hearts of fickle men.

Jesus, the real Jesus of the Bible, is a fierce and proud warrior. Would you worship and submit to Him today?

Resolution

The difficulty in reconciling these truths about God resides in the fact that they all describe the same God. Do not different aspects of your personality govern your existence? It is the same with God. My neighbor across the street sees me with all of our children and has referred to me as the most patient man alive. My wife and my children would attest to a slightly different view concerning the extent of my patience.

The issue is that the love of God is meaningless without the wrath of God. Consider a father who indulges every whim of his sons and fails to discipline them.

Whoever spares the rod hates his son, but he who loves him is diligent to discipline him.
Proverbs 13:24

If you fail to discipline your son, if you spare him the rod, you actually hate your son. Disciplining your son is a supreme act of love. If I only place my hands upon my son in love, he loses something. I will likely raise a self-indulgent heathen. Likewise, if I only place my hands upon my son in anger, in discipline, he loses something. My son will likely grow to hate me. It is only in the full measure of my love and discipline that my sons will grow in faith and when they are older, not depart from it.

It is much the same with God. The love of God is more than seen in His discipline and His wrath. His love is meaningless apart from His wrath. They are different aspects of the totality of His character. They reflect equally, from differing angles, His holiness, His very nature. We cannot neglect either aspect of His character without doing damage to the very idea of God. Let this confrontation with the realness of Jesus drive us to our knees, move us to surrender.

Last Things Concerning Jesus

I will never fully understand anything, much less war, apart from an understanding of our Lord Jesus. Further, I will never practice anything correctly, much less war, apart from our Lord Jesus. Paul, writing to the church at Colossae, says of Jesus,

And he is before all things, and in him all things hold together.
Colossians 1:17

Jesus governs all things, including war. To understand the principles by which he governs, we must understand, as much as possible, the person of Christ. Many people confine Jesus, in their minds at least, to the spiritual realm. If they consider spirituality, they'll acknowledge Jesus, yet the governance of Christ over everyday physicality much less the horrors of war seems abstract, foreign.

War is a decidedly physical endeavor. I seek the destruction of men and things in war. One conclusion we may safely draw upon is that Jesus dealt with the physical nearly as often as the spiritual. He fed and clothed people. He healed them of their illnesses as often as He forgave sin and restored people spiritually.

God calls the believer to imitate Christ, to be continually conformed to

His image. Paul exhorts the Corinthians,

> *Be imitators of me, as I am of Christ.*
> *1 Corinthians 1:11*

Would we not seek as much knowledge as possible of the one we are called to imitate?

As we consider war and turn to orthopraxy, right conduct of war, consider the model given us of the Lord Jesus. Look past the watered-down contemporary caricature of Jesus and see Him for who He is. See Him as the Lion of the Tribe of Judah, the firstborn among all of creation. He is worthy of your praise. He is worthy of you surrender. Of this I assure you.

Chapter 8

8. Centurions and Soldiers

From the Fullness of Time...

On September 2, 31 B.C., the naval fleet of Octavian, under the command of Marcus Agrippa, routed the combined fleet of Mark Antony and Cleopatra. This constituted the Roman Republic's dying gasp and the subsequent inauguration of the Roman Empire.[26] Octavian, the newly declared 'First Citizen', took the name Augustus and declared his reign over the entire known world.

The Empire's ascension ushered in the *Pax Romana* or Roman Peace. None saw fit to challenge Rome's unprecedented dominance for centuries to come. Rome simply had no peer competitors and none would rise. God enslaved the *Pax Romana* to His purposes.

> *But when the fullness of time had come, God sent forth his Son,*
> *born of a woman, born under the law, to redeem those who were*
> *under the law, so that we might receive adoption as sons.*
> *Galatians 4:4-5*

The Pax Romana nurtured a special time, a *fullness*. Creation was ripe for redemption, the conditions set for the advent of the Lord Jesus. A unique, historical epoch, consider the implications with respect to the fullness of time. In general, peace pervaded. Travel and commerce flourished and Christ was born into Palestine, the crossroads of the known world, the hub of the Empire. The *Pax Romana* perfectly set the conditions for the Advent and the subsequent spread of the Gospel message throughout the world.

As the *Pax Romana* defined the existence of the people, consider that

---

[26]Werner Eck, *The Age of Augustus*, (Oxford: Blackwell Publishing, 2003), 38.

the skilled but often brutal hand of the Roman soldier spawned this condition and maintained it. Consider that bloodshed and conquest birthed the perfect conditions for the Advent of the Savior. Roman domination was a fact of life for the common man, reinforced by singular encounters with Roman soldiers, not always the most pleasant of affairs.

The Soldier Before Christ

The Bible offers minimal commentary on the political and military domination of Rome. The Empire serves as a backdrop for the New Testament accounts; Roman soldiers serve as extras. They are part of the scenery, no different in prominence than a tax-collector or a merchant or a fisherman; they just are.

Roman soldiers increase in visibility the closer Christ draws to the cross. All four Gospels record Jesus' arrest in Gethsemane on the night of the Passover though only John records the presence of "a band of soldiers." (John 18:3) The rest refer only to Judas, priests and elders, and a great crowd armed with clubs and swords. (Matthew 26:47, Mark 14:43, Luke 22:47) Were the soldiers there to actually conduct the arrest or just to maintain some semblance of order due to the vigilante nature of the action? As they took Jesus directly to the Caiaphas the high priest and this was a decidedly Jewish affair, the Roman soldiers likely transported Jesus and assisted in His handling. We may safely assume they did not handle Him in the most gentle manner.

Though not a Roman invention, Rome found in crucifixion the perfect punishment. It generated intense agony and provided an exceptionally painful and often protracted death. It also served as a useful deterrent due to the public and humiliating manner of death sending a clear message to everyone involved.

The Jews had another belief concerning crucifixion. From the Law,

And if a man has committed a crime punishable by death and he is put to death, and you hang him on a tree, his body shall not remain all night on the tree, but you shall bury him the same day, for a hanged man is cursed by God.
Deuteronomy 21:22-23

Paul later quotes this verse in describing Christ,

Christ redeemed us from the curse of the law by becoming a curse for us - for it is written 'Cursed is everyone who is hanged on a tree'...
Galatians 3:13

Paul viewed the crucifixion of Christ as more than a simple execution. The Crucifixion fulfilled specific prophecies concerning the Messiah, the Suffering Servant of Isaiah. A simple execution would not suffice, only an execution that 'cursed' the victim would punish in the necessary manner as Jesus literally became a curse on behalf of men. (Galatians 3:13)

The Pharisees seemed to concur as they more than anyone fully realized the implications of crucifixion. Their bloodlust sprang from a desire to demean Jesus to the greatest possible extent as He systematically and increasingly challenged their authority. They lacked the fuller picture of the ironic and prophetic implications concerning the crucifixion of Christ.

Crucifixion served their purposes and so they demanded before Pilate, "Crucify him! Crucify Him! We have no king but Caesar!" (Matthew 27, Mark 15, Luke 23, John 19) In the spirit of concurrence, crucifixion served the purposes of God even more than the Pharisaic objectives. Roman soldiers are the active agents, present throughout in fulfilling prophecy concerning the death of Christ.

Contrary to what many understand, the entire Old Testament looks to Christ, in anticipation of the Messiah who would deliver God's people from their sins. The sharpening and progression of the covenants to numerous specific prophecies concerning Jesus—prophecies concerning his birth, life, and death—the Old Testament exists in anticipation. Not only does the Old Testament speak to His manner of death, it also speaks to His humiliation before men.

Isaiah speaks poignantly concerning the manner of regard for Jesus,

I gave my back to those who strike me, and my cheeks to those who pull out my beard; I hid not my face from disgrace and spitting.
Isaiah 50:6

He was despised and rejected by men; a man of sorrows, and acquainted with grief; and as one from whom men hide their faces he was despised, and we esteemed him not.
Isaiah 53:3

Jesus Himself foretold His treatment as all three synoptic Gospels record Jesus' predictions concerning His arrest, trial, and execution including His treatment and handling during this time. Matthew records the words of Jesus,

And the Son of Man will be delivered over to the chief priests and scribes, and they will condemn him to death and deliver him over to the Gentiles to be mocked and flogged and crucified, and he will be raised on the third day.
Matthew 20:18,20

Mark adds that they will spit on him with Luke recording Jesus' prediction that He will be "shamefully treated." (Mark 10:34, Luke 18:32)

It is Roman soldiers who enable all of this with assistance from the mob and others. Eventually, they assume the primary role as Jesus is transferred from the religious court to the secular Roman judicial system. Throughout this process, Roman soldiers "esteemed him not" in the clearest possible way.

Mark records that an entire battalion, roughly 600 men, participated in the humiliation of Jesus. The rough hands that shoved Jesus to the ground belonged to soldiers. The fists that pummeled his frail frame, bruising and wounding, belonged to soldiers. The saliva that smeared Jesus' blood-streaked face came from the mouth of soldiers. The strong hands bearing the reed that struck Jesus repeatedly in the head belonged to soldiers.

The soldiers fashioned a crown of thorns and jammed it onto Jesus' head. They draped a purple robe over his bony and battered shoulders, kneeled in fake homage, and saluted Him in mockery. They struck him with their hands and spit some more.

"Hail, King of the Jews!"

Soldiers scourged Jesus, a Roman penalty consisting of a severe flogging with a multi-lashed whip containing embedded pieces of bone and metal. Hear the crack of the whip as the soldiers again and again flailed at His bloodied back. Hear His gasps for breath at the astonishment of the sharpness of the pain, the bone and metal ripping and tearing flesh and bone. Consider that the blood of Christ first spattered across the face of a Roman soldier. They spit some more and led Him again before Pilate. (Mark 15:16-20, John 19:1-4)

"Behold the man!" Pilate declares. (John 19:5)

"Crucify Him!"

Pilate relents to their demands and sends him off to be crucified. Soldiers lead Him to His fate, forcing Jesus, in His weakness, to bear the instrument of His death, His own cross on the dreadful march to Golgotha. The soldiers continue with their barbarous treatment during the awful procession.

Upon arrival at Calvary, consider an often neglected aspect of the humiliation of Christ, that the soldiers likely ripped Jesus' clothes from Him, crucifying Him in the shame of his nakedness. In Jewish culture, viewing the genitalia was considered decidedly offensive. Rome normally crucified men naked, compounding physical agony with shame and humiliation. Though not specifically recounted in the Bible, and though almost all depictions in history show Jesus wearing a tattered loincloth concealing his genitals, the soldiers likely stripped him bare. The shame of the exposure of his genitals, the vulnerability, coupled with His tattered flesh and the derision of the soldiers fully completed Jesus' humiliation.

Soldiers, in their strength, drove Jesus to the ground at Golgotha. Strong arms pinned His frail limbs to the rough wood of the cross beam. The sweat of their brow, in the exertion of their labor, dripped onto Jesus' writhing frame as the executioners pinned His arms and positioned the nails.

Did one hold a spare nail in his mouth in the fashion of a man doing a bit of light carpentry around his home? Did one ask the other to hold Jesus' arm steady? Did an overseer offer bits of wisdom?

Steady.

A little higher.

Hold 'em.

Angle it a bit more.

When they finished driving the nails, soldiers raised the cross, lifting Jesus, the Son of Man, as Moses lifted the serpent in the wilderness. (John 3:14) The soldiers then cast lots for Jesus' tunic (John 19:23,24) so close to the cross, yet so far from Christ, and mocked Him yet again, offering Him sour wine to drink. (Matthew 26:34)

"If you are the King of the Jews, save yourself!" (Luke 23:36)

The Jews appealed to Pilate to break Jesus' legs that He might die sooner from suffocation, unable to support His bodyweight. They didn't want the execution to drag over into the Sabbath. A soldier wielded the iron rod, swinging fiercely and connecting with a loud report, shattering the femur of first one and then the other criminal. Coming to Jesus, the soldier noticed He was already dead and so another soldier pierced His side with a spear, thrusting its sharpened tip deep into Jesus' ribs. Blood and water sprayed from the wound, dousing the face of the soldier. (John 19:31-37)

The face of a Roman soldier was quite possibly the last face that Jesus saw before He closed His eyes and gave up His spirit in death. Even in death, Roman soldiers guarded His tomb.

Following the Resurrection and Ascension, Christians increasingly ran afoul of Roman law and religion as the Church exploded around the Roman world. Nero, Vespasian, Titus, and Domitian oversaw an increasing general persecution of Christians. Roman soldiers crucified Christians including the Apostle Peter. Roman soldiers beheaded Christians including the Apostle Paul. They mercilessly beat Christians, seeking recantation. Others thrust Christians into the arena before wild animals and gladiators, satiating the blood lusts of the rabid crowds.

The year A.D. 70 spawned the culmination of Roman military barbarity as Titus lay siege to Jerusalem and slaughtered the Jews, desecrated the temple, and destroyed the city in a time of unprecedented carnage and tribulation, a time foretold by Jesus in Matthew chapter 24. Josephus wrote that, "Round the Altar the heaps of corpses grew higher and higher, while down the Sanctuary steps poured a river of blood."[27]

From Jesus' birth to His death, Roman soldiers are there. They carried out the Slaughter of the Innocents recorded in Matthew Two whereby Herod, out of sheer paranoia, demanded the death of every male child in Bethlehem under the age of two. Only by His parent's escape to Egypt did Jesus avoid death at the hand of Roman soldiers. Thirty years later, Roman soldiers finally accomplished that which they'd been ordered to complete so many years before, killing Jesus. Roman soldiers, and all soldiers by extrapolation, occupy a surprising position in the heart of the One they sought after and destroyed.

The Soldier unto Christ, a Reconsideration

I'll not claim to know the mind of Christ. I merely state that which Scripture reveals and then offer my own conjecture, colored though it may be. As we've seen concerning the treatment of Christ at the hands of the soldier, concurrence first offers relief. Luke records Peter's sermon at

---

[27] Peter Schafer, *The History of the Jews in Antiquity*, (Routledge: New York, 1995), 191-192.

Pentecost,

> *this Jesus, delivered up according to the definite plan and foreknowledge of God, you crucified and killed by the hands of lawless men.*
> *Acts 2:23*

He records the later testimony of the brothers,

> *for truly in this city there were gathered together against your holy servant Jesus, whom you anointed, both Herod and Pontius Pilate, along with the Gentiles and the peoples of Israel, to do whatever your hand and your plan had predestined to take place.*
> *Acts 4:27-28*

The Crucifixion served the "definite plan and foreknowledge of God". It accomplished what God "had predestined to take place." Though the Father ordained the Crucifixion, the hands of lawless men actually accomplished the most evil act in the history of the world, according to their own evil will and intentions. Their evil wills served God. What they intended for bad, God intended for good, the greatest good that has ever been devised.

Peter never condemns the actual hands of execution, those of the soldier. No, he speaks against the lawless men who drove the act. Without them, the soldiers never would have executed Jesus. Without Herod, the Pharisees, without Pilate, the Jewish people, without the sinful hearts of wicked men, the soldiers never would have received the order to execute Jesus. The lawless men refers to a broad swath, from the individuals mentioned to the collective wicked hearts of all humanity.

At Pentecost, Peter preached to "Men of Israel," (v.22) and declared "you crucified" Him. (v.23) "You crucified Jesus," Peter preaches to the men of Israel. Extrapolating this in light of the Church returns a broader and more troubling answer. Who crucified Jesus? You, believer. You crucified Jesus. As the hymnist sings,

> *Behold the man upon the cross;*
> *My sin upon his shoulders.*
> *Ashamed I hear my mocking voice,*
> *Call out among the scoffers.*[28]

You, believer. You scourged Jesus. You mocked Jesus. You spit upon

---

[28] Stuart Townend, "How Deep the Father's Love For Us."

Jesus. You betrayed Jesus. Yes, your sin put Him on the cross. You killed Jesus.

Not the soldier.

To be sure, we are not yet considering redemption. As the believer's sin drove Jesus to the cross, the ultimate answer to the query of who killed Jesus is God. The Father slew the Son on behalf of those who would believe. Judicial guilt generates life. Implication in the death of Christ, having your sin laid upon His shoulders, effectively eliminates future and eternal judgment, thus the nature of the Atonement. The point to this line of reasoning is that nowhere does God condemn the soldier, writ large, for his role in the murder of Jesus.

An examination of Scripture proves useful in establishing the very opposite.

Cast Aside Your Weapon

The Bible speaks concerning itself,

All Scripture is breathed out (inspired) by God and profitable for teaching, for reproof, for correction, and for training in righteousness, that the man of God may be complete, equipped for every good work.
2 Timothy 3:16-17

Every word of Scripture is entirely intentional. Coupled with sovereignty, we may exposit that no accidents reside in the Bible any more than they do in life.

It is no accident that you have picked up this work any more than it was accidental that Peter, at Gethsemane, cut off the *right* ear of Malchus, the high priest's servant. Why the right ear? Why not the left ear? Why not his nose or maybe his hand? Some things we'll never know this side of glory. Perhaps the significance resided with Malchus for some unrevealed reason. Rest assured, a reason exists, unrevealed though it may be.

At the Crucifixion, a criminal and a soldier first recognized Jesus as the Christ…for a reason.

As Jesus hung dying, He cried out with a loud voice, "Father, into your hands I commit my spirit!" (Luke 23:46) The curtain, the veil of the temple that separated the Holy Place from the Holy of Holies, was torn in two, from top to bottom. The earth shook, rocks split. The tombs opened and many of the bodies of dead saints were raised and came into the city and appeared to many. (Matthew 27:51-53) Matthew records that,

When the centurion and those who were with him, keeping watch over Jesus, saw the earthquake and what took place, they were filled with awe and said, "Truly this was the Son of God!"
Matthew 27:54

As Luke records,

Now when the centurion saw what had taken place, he praised God saying, "Certainly this man was innocent!"
Luke 23:47

As Jesus breathed His last, He possibly beheld a soldier. It is a centurion, a soldier, who first declared from among the masses, "Truly this was the Son of God!"

"There is therefore now no condemnation for those who are in Christ Jesus," (Romans 8:1) rings as true for the soldier as it does for the accountant, the tax collector as much as the preacher. Condemnation or redemption reflects one's position in Christ, not one's occupation or profession.

I can think of no legitimate occupation that merits automatic judgment. Some would consider the soldier as one automatically condemned based upon misguided perceptions about God. I believe that some soldiers hold similarly misguided views. I find no call for this in Scripture.

At the same time, I know others who so esteem the occupation of soldiering that they refute the notion of accountability along the lines of, "He (the soldier) has served his time in hell." Again, I find no scriptural basis for allowance or leniency for any occupation—be it a soldier or a priest.

Again, our position before God is a function of our position in Christ. Perhaps Jesus and John the Baptist, in earlier encounters with soldiers, might help clarify this point.

An Account to Consider

Matthew chapter eight records that as Jesus entered Capernaum early in His ministry, a centurion came to Him.

"Lord, my servant is lying paralyzed at home, suffering terribly." (v.6)

"I will come and heal him," (v.7) Jesus responded to which the centurion replied,

"Lord, I am not worthy to have you come under my roof, but only say the word, and my servant will be healed. For I too am a man under authority, with soldiers under me. And I say to one, 'Go,' and he goes, and to another, 'Come,' and he comes, and to my servant, 'Do this,' and he does it." (v.8,9)

Matthew records that Jesus "marveled" at the centurion's response. (v.10) He declares to His followers, "with no one in Israel have I found such faith." (v.10)

Then, Jesus pronounced judgment on non-believers and to the centurion, "Go; let it be done for you as you have believed." (v.13) The servant was healed at that very moment.

At this point, Jesus had already called the brothers Simon Peter and Andrew and James and John to follow Him and likely the rest of the Apostles. He had preached the Sermon on the Mount, healed many, cast out demons and raised a young girl from the dead. Many followed Him in faith. Some in fact, left every aspect of their lives to follow Him and yet, He declares of the centurion, "With no one in Israel have I found such faith." (v.10)

Jesus marveled at the faith of this fighting man instead of castigating him for pursuing the profession of arms. Jesus never once even mentions His occupation. In touting his faith, Jesus doesn't say, "Despite being a soldier, look at this man's faith." No, He extols his faith for what it is, independent of his legitimate occupation.

Telling is the centurion's posture before Jesus. In humility, he declares himself unworthy to even receive Jesus into his home. He understands authority and further understands Jesus as one whose authority usurps all, that by His word He can heal. Consider that this man of some means came to Jesus to appeal on behalf of a lowly servant, indicative of the condition of his heart. Again, his position in Christ drives his position before God, even as a soldier.

Luke records in chapter three of his gospel an early encounter between John the Baptist and several soldiers. John is preaching and teaching in the wilderness, baptizing those who come and pronouncing judgment upon the wicked, particularly the evil of the religious establishment. In confusion, the crowds ask him, "What then shall we do?" (v.10) to which he urges them to share what they have. Tax collectors ask the same to which he exhorts them to, "collect no more than you are authorized to do." (v.12)

Soldiers among the crowd likewise enquire, "and we, what shall we do?" The soldiers clearly misunderstand their position before God as a soldier. I can see the line of reasoning. I'm a soldier. I spill blood for a living. What should I do? These other folks can do what you say, but what about us? How shall we avoid condemnation?

John advises them, "Do not extort money from anyone by threats or by false accusation, and be content with your wages." (v.14) In other words, conduct yourself honorably as a soldier. Don't abuse your position. Coupled with John's exhortation to the tax collectors, I could hermeneutically advise a soldier to "spill no more blood than you have to." Take no more life than you have to in the fulfillment of your duty. John never once calls them to lay down their swords. Neither does Jesus.

One further account will prove useful in establishing a decisive and coherent biblical view concerning the soldier.

The early church was a decidedly Jewish institution. The Apostles were Jews. All of the first converts were Jews. Jesus Himself was a Jew of the tribe of Judah. The possibility of Gentile salvation cut across the grain for early Jews as their contempt of all things concerning the Gentiles resonated throughout their collective psyche. They considered Gentiles ceremonially unclean and impure. Gentile lands were ceremonially unclean to the point where if a Jew even walked in the Gentile lands, he was to shake the dust off of his sandals prior to walking upon the blessed dirt of Israel.

God intended salvation for men from every tribe, tongue and nation. (Zechariah 8:23, Revelation 7:8,9) Unlike the Jews, God is no respecter of persons. Jewish racial hostility stemmed from a history of Gentile domination coupled with an elitist attitude developed from sheer religiosity. That the promise of Abraham extended to all the nations was soundly ignored or suppressed. Early Christians struggled with their Jewish preconceptions.

God's ways are perfect and He selected a soldier as the first Gentile member of the Church. As Acts chapter two records the founding of the Church at Pentecost, Luke records in Acts chapter 10 the salvation of Cornelius, a centurion. Cornelius is stationed in the coastal town of Caesarea and is a member of the Italian Cohort, an unknown Roman army unit of probably about a 1000 men.

Luke describes Cornelius as, "a devout man who feared God with all his household" and that he "gave alms generously to the people, and prayed continually to God." (Acts 10:2) Unlike many Roman soldiers, Cornelius treated the people, the common man, with kindness indicating the condition of his heart. He worshiped and pursued the one true God as the Lord had already tilled the rocky soil of his heart.

An angel appears to him in a vision one particular day as he prays. As directed by the angel, Cornelius sends for Peter at Joppa. Peter has a vision the next day, as Cornelius' men journey to Joppa. As his vision ends, Cornelius' men from Caesarea knock on his door. (v.17) Peter returns with the men to Caesarea the next day.

Cornelius, expecting Peter, falls to his feet to worship him to which Peter responds lifting him up, "Stand up; I too am a man." (v.26) They talk for a few minutes and Cornelius relays the content of his vision from a few days prior. As he had gathered his relatives and close friends, Cornelius says, "Now therefore we are all here in the presence of God to hear all that you have been commanded by the Lord." (v.33)

Luke records that, "Peter opened his mouth and said:

> *Truly I understand that God shows no partiality, but in every nation anyone who fears him and does what is right is acceptable to him. As for the word that he sent to Israel, preaching good news of peace through Jesus Christ (he is Lord of all), you yourselves know what happened throughout all Judea...and he commanded us to preach to the people and to testify that he is the one appointed by God to be judge of the living and the dead. To him all the prophets bear witness that everyone who believes in him receives forgiveness of sins through his name."*
> Acts 10:34-43

What a day this was! God chose this day to bring salvation to you the Gentile and to you, the soldier. Why did God choose a Roman soldier as the first non-Jewish recipient of the Gospel?

Perhaps He understood the intense vulnerabilities in the psyche of the soldier, the vulnerability to misunderstanding, to misinterpreting their position before God in terms of their occupation. Recall that Scripture contains nothing incidental. No other legitimate occupation bears a similar burden, largely due to the inherent difficulties in the destruction of life and active participation in such destruction. God understands this and so he dispatched Peter to proclaim the Good News of salvation in Christ Jesus to Cornelius the centurion. Notice Peter's opening words, "God shows no partiality." (v.34)

The idea of diversity, of the universal nature of the Church, not universal in the sense of a totality of inclusion but universal in terms of representation, runs contrary to sinful human thought. Surely my position bears special consideration. Surely I am excessively unworthy because of my occupation. Surely I am excessively unworthy because of what I've done. Surely I am unworthy because of who I am.

That is the beauty of the Gospel. A lack of worth permeates all of humanity. In jest but ringing true, Gunnery Sergeant Hartman in the movie *Full Metal Jacket* derides his trainees, "Here you are all equally worthless!" All men stand, in equality, equally condemned before a holy God. All men stand with but one hope, which is salvation in our Lord Jesus. Salvation is available to all who would believe, no matter your occupation, no matter your previous transgressions.

Clearly God previously tilled the rocky soil of Cornelius's heart such that, upon hearing the Gospel for the first time, "the Holy Spirit fell on all who heard the word" (v.44) and Cornelius and those with him were saved. Peter baptized them and reported back to the church in Jerusalem and "they glorified God, saying, 'Then to the Gentiles also God has granted repentance that leads to life.'" (Acts 11:18)

At this, God inaugurated the completion of His promise to Abraham, to bless all the nations through Israel. He chose a soldier as the vehicle for this fulfillment.

A Final Regard

Consider once more the intentionality of Scripture. Each word is completely intentional, meant to convey an intended meaning to an intended audience for an intended purpose. We may declare with authority a number of conclusions pertaining to God's regard for the soldier.

1. God issues no special condemnation for the soldier.

2. God calls the soldier to conduct his duties in an honorable manner. (We'll discuss the implications of this in a subsequent section)

3. God maintains a certain regard for the soldier.

4. God's regard for the soldier merits no special consideration with respect to accountability in light of salvation.

As all men suffer, the soldier suffers uniquely in light of the distinctive nature of soldiering, of the necessary duties imposed herein concerning the destruction of life. Soldiering demands strength and courage due to the difficult task at hand as well as the potential for damaged perception in light of a misunderstood positioning before God with respect to occupation.

Indeed, the flawed aspects of sinful human cognition coupled with popular tendencies to misportray place a notable burden upon the soldier.

Thus, when Paul exhorts Timothy to "Share in suffering as a good soldier of Christ Jesus," (2 Timothy 2:3) he declares simultaneously the burden of soldiering and the inherent potential goodness in the soldier. Why else would he tie these two ideas together, suffering and the ability to bear it well? We'll address this burden and the necessity to bear it well shortly.

Consider your call and the freedom in knowing that the God of the Universe esteems the honorable soldier. Rejoice in this knowledge, revel in it as we turn a deaf ear toward all that falsely claim otherwise. Recall Peter's words to Cornelius, "God shows no partiality," and glorify God in this truth.

9. Saved Rounds

God saw fit to test me sooner than I anticipated immediately following my conversion in 2005.

Once the war started, I deployed fairly regularly for a number of years until my return from a particular trip late in 2004 when my commander moved me to battalion staff. He needed a trusted agent NOT to deploy, to serve as the operations officer and run the battalion. This meant stewing in the rear and dispatching others to do the fighting, not something to which I endeavored. Yet, I poured myself into the work, determined to honor my boss, unknowing at the time that God had ordained this time of tranquility to facilitate His call.

Early Life

After conversion, Ami and I joined the church, got baptized, and began attending a Sunday school class. I still remember how surprised I was to learn that these people were just as messed up as I was. These people were sinners! I guess I had been expecting a church full of perfect people yet what I found was a church full of sinners saved by grace. I made some amazing friends and my newfound faith began to grow. Eventually my personal *Pax Romana* came to an end as the unit needed me forward in combat once more.

My first deployment following conversion was a warm-up of sorts, a staff rotation. I would be the operations officer for a task force in Iraq, planning combat, but not actually conducting combat. I had not yet reconciled my faith with warfare so I went in asking lots of questions, praying, and reading. I watched myself. How would this affect me? What would it do to me?

What I found was that after the rotation I felt...absolutely nothing, nothing more than satisfaction at a job well done. The operational tempo had been fairly intense as we executed three to four operations a night, dispatching hordes of bad guys. I felt nothing. Maybe it was because I was so far removed from the action, tucked safely away in the operations center. Still I monitored myself.

At some point, the unit established another task force that I was qualified to command were I to get a quick aircraft transition. Sensing an opportunity to get out of the operations center and into the cockpit, I finagled a slot in a three-week aircraft course and maneuvered to get on the deployment tracker, offering myself up to command the new task force.

The headquarters relented and I was able to deploy again in command of this new task force, this time in the cockpit actually executing combat. The pace of operations remained intense as we executed several objectives each night and still I observed myself and still…nothing.

Living a Dream

At one point I was granted my dream job, command of a gun company. Though I loved putting the assaulters on the objective, I longed to be a gun pilot for as long as I could remember and after years of maneuvering was granted the job. For two amazing years I got to pretend to be notorious. I absolutely loved it, relishing each day. It actually ruined me from flying for when I finally gave up the command, I had no desire to fly. If I wasn't shooting, I had no interest. I forced myself for a couple of years but at some point, I just gave it up. I'd lost the taste.

During this command, I entered the most heavily deployed time of my career—a whirlwind of deploy, recover and train, deploy again. My first deployment as a gun pilot brought me to the ancient city of Ninevah. Like before, we executed at least one mission a night and like before, I watched myself closely.

One engagement highlighted the non-issue. So well-practiced were our tactics, so effective were our procedures, so skilled were our customers, that the bad guys almost always surrendered without a fight or often were subdued or killed before being able to muster even a shot. They rarely stood a chance and we always sought to stack the deck in our favor as much as possible. We always sought an unfair fight.

Imagine being the enemy, if you will. You're out cold on your sleeping mat, the murmuring of cattle and donkeys within the walls of your compound soothe the night, a dog barks in the distance. Your wife(wives) lie scattered around the room, children tucked into every nook. The brisk night desert air chills any exposed body parts.

Suddenly, your heart leaps right out of your chest, rocking you to the core of your fiber as the wall of your home explodes inward showering everyone with dust and debris. In shock, you gasp for breath, struggle to hear. Your entire home vibrates, alive with movement. A whirling dervish, a maelstrom of dust, engulfs the courtyard of your home. Rough footsteps, more explosions, silent figures flirt with the shadows.

Green eyes set upon you from the darkness. You become frozen, incapable of motion. More green eyes as children begin to scream and before you can fathom what is happening, something impossibly solid strikes you across the jaw, jarring your conscience, shattering your reality. In an instant, you find yourself face down in the dirt, pressed to suffocation, roughly bound as the screaming of the women and children punctuates the night. Mere seconds separate peace from bondage, possibly death.

A Practiced Affair

I'd actually been on the receiving end once. My friend Bobby came over to the unit from several years as a ground guy. We were on an urban training event and I had the night off. The training chief for the boys had been a friend of Bobby's and so he asked Bobby and me if we'd like to be on the objective as the assault went down. I was more than happy to accept the offer.

The objective was a three story building, an abandoned hospital and as Bobby and I arrived, we could see all of the role players getting ready. The assault would happen in a few hours and the pretend bad guys were busy gearing up. A chem light hanging around your neck on the objective meant that you were out of bounds, off limits, merely an observer. Expecting a chem light, the chief pointed to a red man suit and told me to get dressed.

I hesitated, "uh what?"

"I need a role player, Bobby said you were up for it."

I looked at Bobby who smiled trying to look innocent.

"All right," I conceded reluctantly and proceeded to don the padded suit and helmet. Once dressed, the chief led me to a vacant room on the third floor.

"Okay, here's the deal. When the assault happens, these two dudes here will engage them for a few seconds and then move out. You'll stay here."

"Do I get a weapon," I inquired.

"No, you're an unarmed combatant," he replied as I was not quite yet comprehending the implications of what was transpiring. "When you see 'em, get 'em," were his matter-of-fact instructions to me.

I looked at him hesitantly. "Okay, what exactly do you mean, 'get 'em'?"

He asked me if I played football and I informed him that I had.

"When you see the first guy, put him on his ass!" he instructed me.

"Okay, yeah but…how hard. What do you mean?" still very uncertain.

"Take them the #*&@ down! But realize, they'll do what they have to do to subdue you," and with that, he was off, leaving me in a cloud of uncertainty.

I was about to go head-to-head with the boys! Adrenaline coursed through my veins, my heart rate skyrocketed, my muscles tensed. H-hour, the initiation of the assault, was only an hour or so away.

I paced that room like a caged lion, impatiently awaiting my demise. I must say, being on the receiving end of an assault was an eye-opening experience, and I knew it was coming and when.

As the hour approached, I became even more intense, more focused. My breathing began to fog up the faceplate of my helmet, obscuring my vision. I took my helmet off and replaced it just as quickly. Two minutes out. There it was, a high pitched hum that roared to a crescendo, infil to the roof! Right on time. The muffled boom of the breach followed by the entire building quaking with vibration as the main body maneuvered over the roof. Voices shouting, another breach.

The two bad guys in the adjacent room began to fire down the hall at the yet-to-be-seen assault force. After a few shots, they maneuvered away, leaving me alone. This was it. My fists balled as I crouched, ready to pounce. More shouts, gunfire, another breach. Let's do it!

Suddenly, I noticed motion out of the corner of my eye from a direction I wasn't expecting. The boys were quick, flowing through the objective like water. With a shout, I steeled my nerves and charged, determined to absolutely destroy the first dude.

The next few seconds were a blur. I was buffeted, all I remember, slammed to the ground as I desperately grabbed for weapons, equipment, appendages. Quickly a knee drove into my back, pinning me to the ground, maybe more than one knee. Arms secured my arms. A strong arm encircled my throat and practically twisted my head right off my neck. I was done. I tapped the strong arm in surrender. More twisting, choking now. I tapped harder.

A voice, "Hey dude, he's tapping."

"I know", to me, "you done?"

"Yep, yep," I managed with a choke.

With that, they zip-tied my hands behind my back and I lay awaiting end of mission. It wasn't long before the chief came and freed me, helping me to my feet.

"How'd it go?" he asked.

I looked to see a smiling Bobby standing nearby. What a friend! "It was awesome!" I responded.

"Holy crap dude, they got you good," Bobby remarked pulling my shirt collar down to reveal an angry red circle on my neck, courtesy of a brilliantly executed muzzle tap. I hadn't even felt it at the time. I will definitely never forget being the target

Ready to Die

On this particular assault out of Ninevah, the bad guys seemed almost ready. Dub and I maneuvered our gunship as dash two over the objective and waited. The ground force surrounded an isolated compound that contained one of the countless nameless bad guys we'd engage. At that point, as it was the middle of the night, they chose to go with the loudspeaker vice a hard breach, literally a, "Come out with your hands up, we've got you surrounded."

We watched from about a mile away. I was on the sensor watching the action in real time, keeping track of where the good guys were, ready to roll in hot if things turned south. The bad guys needed some extra encouragement this night.

After a couple of chances and a reminder of the futility of resistance, the boys offered up some encouragement by way of a volley of machine gun fire into the windows. Still nothing. They needed a bit more encouragement so the boys delivered a thermobaric LAW into the window of the hut. The explosion temporarily blinded my sensor and still I watched and still, nothing. These dudes really needed some encouragement.

I watched as the boys tightened up the perimeter a bit and listened to their comms as they discussed a couple of COA's. Finally, the ground force commander decided to give thermobarics one more shot before commencing an assault. This time, they fired three simultaneous rounds through three separate windows. The explosion was magnificent. As the smoke cleared and my sensor regained resolution, I was surprised to see the building still standing. Still I watched and finally, movement.

Slowly, two figures emerged onto the roof of the building, hands in the air in surrender. Thirty or so rifle barrels zeroed in on their skulls. They inched toward the edge of the building, hands still raised… when it happened. Their hands dropped as one of them flung an object at the ground force. An instant volley of fire raked the rooftop as the assaulters leaped for cover. Grenade! One of the bad guys had decided to go down fighting.

The grenade exploded harmlessly and we moved in. I could still see the two bad guys laying on the roof, moving around a bit, wounded but still very much alive and still very dangerous. The call came, "hellfire on building one."

"Roger, we're inbound, ETA 1 min."

We quickly sweetened up our lasers and Dub asked, "You got this one?"

My first gunship engagement, "I got it."

We would laze the target for our lead's missile. I zoomed in my sensor and held the crosshairs about a third of the way up the side of the building just as I'd been taught. The two bad guys continued to move around on the roof, knowing their time was short, but not knowing from where death might come.

"Spot on." Lead was tracking, in constraints. Fire. Missile off the rail. Riding our beam to the target. I held the crosshairs.

"Steady, steady," Dub soothed in my headset.

After an eternal few seconds, impact. The missile rode true delivering its payload of death to these two wayward men in an instant, vaporizing them along with much of the building. We maneuvered for a second shot but it would not be necessary. As the dust and smoke cleared a bit, my sensor revealed a pile of rubble where the building had once stood.

"Good effects," from the controller. With the threat eliminated, the assaulters were confronted with the unenviable task of sifting through the rubble for bodies and any other intelligence they could find—computers, cell phones, documents, anything and everything that might lead to another bad guy.

I marveled at the brazenness of these two men. Had they gone to sleep this night anticipating martyrdom? They hadn't known we would show up, but when we did, without hesitation, they were ready to fight to the death, an objective we were more than happy to assist them with.

I had a morbid but amusing thought. Had both of them been on the same plan or had one actually been planning on surrendering, unwitting to his comrade's intentions until he threw the grenade. Were they on the roof mortally wounded with one of them gasping, "You idiot! What are you doing?!"

Dub and I flew in silence back to the staging base. All in a night's work. Again, I examined myself and again…nothing. As had become my custom, I prayed following the engagement. I prayed for the soul of these two dead men. I prayed that God would have mercy on them. I prayed that somehow, in the instantaneous violence of their demise, God had miraculously saved their souls and I grieved at the thought of these two men, closing their eyes in death, and opening them to eternal torment, expecting virgins, finding hellfire.

I did all that I knew to do at the time; I kept praying.

Section 2:
War and the Mind

Chapter 10

Tuesday, May 3rd (30,000 ft above the CONUS)

There exists in my unit a peculiar phenomenon that must exist elsewhere, though in all fairness I've only observed it within these ranks.

I'm not sure exactly how to label this particular phenomenon, but we'll start by calling it *the point of culminating preparation*. I am writing this particular excerpt from 30,000 feet above the United States in the belly of an Air Force C-17 aircraft, a monstrous beast capable of hauling hundreds of men and mountains of equipment back and forth around the globe and without which, the war effort that I've been a part of for over a decade would certainly have languished at some point in time. I simply cannot remember how many trips I've made aboard one of these marvelous machines.

The switch has been flipped. I awoke this morning with a simultaneous sense of dread and urgency. I'd been living a duality for several weeks, dreading the arrival of this day yet wanting to get on with it as rapidly as possible so I could punch the clock once more and start the countdown. Predictably, these last several weeks flew by, almost as if they never existed. Though I'd enjoyed a relatively luxurious preparation timeline as I truly had no other competing professional demands, I still found myself rushing a bit, forgetting this tidbit or that little piece of equipment, right up until the end. Even then, it's not complete. I'm sure I've forgotten something.

As arduous as combat may be on occasion, the littlest of things can drive that which we all seek, a bit of comfort here and there. I forgot my shower shoes once. I don't remember where I was or what I was doing but I remember quite vividly the extremely laborious process of putting my boots on to walk to the shower tent, taking a shower while walking on the absolutely tiniest bit of the outside of each foot so as to minimize the amount of my flesh that came into contact with the disgusting germ-ridden shower floors, and then putting my boots back on to trudge back to my tent, only to take them off to don the rest of my uniform. Arduous!

As it is, the weeks preceding a deployment lead up to a singular point, a threshold, this phenomenon that I seek to describe—*the point of culminating preparation.*

I showed on time this morning and dropped off my tough box for the pallet build and then met Ami and my people for a quick lunch. Thus far, things had gone relatively smooth. I'd alerted all of my people in a timely manner and then on Sunday night after church, I assembled the family in the living room for a pre-deployment brief.

After discussing the week and reminding them of my impending deployment, the location, my mission (to fight the bad guys of course), and the duration, I reminded them all how much I loved them and that I expected them to be on their best behavior and to help their mom out.

Our family dynamics changed a bit since my last deployment. Then, our modest tribe included our three daughters and two sons. Now, though my oldest daughter and son had moved out on their own, we'd added four more sons leaving my poor wife home with our two daughters and, gasp, five boys…five very rough and rowdy boys. Just the other night I locked my 15-year-old into a wicked rear choke just to remind him of the pecking order, something I feel is necessary now and again. Now, for a number of months, he would receive no reminder. I prayed for tranquility, for peace at home.

Each night this week, as I tucked the little guys in, we talked more about the deployment. Last night was the hardest. I love to watch my little guys sleep. I'm not sure if all parents do, but I do. I relish it. After I read them their story and we said our prayers, they drifted off to sleep and I stood looking at them. Really, for the first time, I began to feel the pangs of separation. With quiet confidence, not over-confidence, I was and am fairly certain of my return, but I will miss them greatly. I gently kissed each of them on the forehead and squeezed my cheek against theirs. DJ, in his raspy half-sleepy voice, "I wuv you daddy."

"I love you too, son," as I kissed his cheek one last time.

I went upstairs to where Ami was lying on the couch, cuddled up with baby Max. I sat next to her in silence for a few minutes looking at the baby and again, the pangs of separation began to eat at me. Max had come to live with us when he was two months old and at 14 months, it looks like he is here to stay. I am his father now and his eyes light up when he sees me. No matter who is holding him, except for maybe Ami, he reaches for me as soon as he sees me.

When I toss him in the air, he giggles uncontrollably, probably the most joyous sound I know. When I hide from him and then pop out, he squawks at me and turns to run but always looks back to ensure I am coming to get him. I'd forgotten how hard it is to leave a little one.

With a lump in my throat, I asked Ami to please be sure and remind him about me. I began to well up a bit and Ami, as she always does, delivered a pep talk.

"Baby don't, these boys need to see this. This is hard, but it's so good for our family. The boys need to see their daddy going to fight the bad guys, being a hero. They've never had that. They don't even know or understand this or what sacrifice is."

All of my sons came to us via adoption, but two of them we recently adopted out of horrifically awful circumstances complete with a drug-ridden, absentee father. Just last week he was implicated in the death of their oldest biological sibling, essentially shooting him up with drugs when he has a known heart condition. At some point, their brother went to sleep and never woke up.

I could see her logic, but it didn't replace the hole in my heart or soothe the dread in my gut. Tomorrow was the day.

The next morning, this morning, I got up early, got in a quick lift (gotta stay pumped) and headed into the compound before meeting Ami and the people for lunch. Subway was good and uncrowded and we had a peaceful and uneventful lunch. We lingered a bit, but after a few minutes of chatter, I looked at Ami. It was time.

"You ready, baby."

"I guess so."

I carried baby Max to the van, squeezing him tight as if I'd never squeeze him again. I hugged my youngest daughter and then my 15-year-old son, imploring them once more to help their mother, reminding them that I love them. Juan Miguel gave me a big squeeze and DJ hugged me tightly. Even my 12 year old leaned into me and told me he loved me. Ami always got the last and the longest, squeezing her face into my neck, stretching the moment, not wanting it to end, yet knowing it must. Her tears wet my neck a bit, just a bit.

"Be careful baby."

"I will," and with that, I climbed into my truck, looked back at her one last time, and drove away only to sit at the traffic light and sob quietly for a minute, just a minute. Then, the switch flipped. Time for work. I could almost put my guard down, relax, and get busy with this job, almost, but not yet.

Back to the phenomenon. I had not yet reached *the point of culminating preparation.*

As I write from 30,000 feet, I am literally sprawled across the floor on top of my sleeping mat. I've shed my uniform outer shirt and even my pants, trading them for a comfortable pair of black shorts. My head rests just beneath the chin bubble of a helicopter and I have enough space on either side of me for at least two or three men. After writing this part, with Benadryl singing me to sleep, I'll stretch out, cover up and sleep, hopefully for the duration of the flight though the gyrations of the in-flight refueling will likely wake me up in a few hours.

On each side of the helicopter, another 40 or so men lie in similar states. Many are already asleep. A few watch movies on their portable electronic devices. One odd soul is actually still in his seat, reading a book of all things. This is the phenomenon or rather, it is the achievement of this state, that all of these men seek, have been seeking for some time.

Preparation for combat is complete. They've trained. They've prepared themselves emotionally, physically, mentally, financially even. Like me, they've left behind families, wives, and children. As the looming deployment approached, their preparation narrowed to a singular event, a singular focus—occupying the most perfect piece of real estate on the floor of the C-17 and setting up camp.

As minuscule and trivial as this may sound, I've noticed repeatedly, every time, the intense amount of jockeying that happens concerning positioning and the lengths to which some go to not only obtain the perfect piece of real estate, but to ensure its suitability.

The helicopter flight crews always arrive at the C-17 first and immediately they hang their jackets on seats, claiming them for themselves and their friends. Usually, they know the content of the load and therefore know which strategically sound location to choose—never by the fuel tanks of the aircraft or God forbid, the refueling probe as a healthy and prolonged whiff of jet fuel will be your fate, and never by one of the aircraft tie-down chains as they severely restrict your ability to pronate.

The crews leave the poor staff and support guys to fend for themselves as they have to wait for the helicopter to be loaded before coming aboard. Once manifest call is complete, they storm aboard, searching for a good seat, one with enough adjacent room to go horizontal. Frantically they search while the helicopter crew members sit smugly in their strategically selected seats, confident in their imminent 9 hours of uninterrupted rest.

Some notorious sections dispatched their new guys to the load early, merely to claim seats, and God help the young pilot who failed to secure a suitable seat for the senior pilots. Once the C-17 takes off, all of the soldiers sit in utter anticipation of the moment when the Air Force loadmaster announces that they are free to unbuckle their seat belts and occupy the floor. You can almost hear the audible sigh, the relief.

Not all trips are as luxurious as my current one. I deployed once with the Ranger Regiment, packed in like sardines. They occupied the floor all right but with several hundred men on board I found myself with young Rangers pressed tightly against me on each side. They were under things, over things, around things, and most of all, around me. I could not wait for that flight to end.

This flight, with a very reasonable 40 or so soldiers, offered ample room for all, greatly easing everyone's angst. On the crowded flights I'd been on before, tactics became quite cutthroat when jockeying for position.

Over the years, the lengths to which the soldiers have gone to enhance their inflight experience has become increasingly extreme. Several years ago, I witnessed a soldier unfold a nearly twin-sized Posture-pedic mattress of all things to ensure he received adequate in-flight rest. Some roll out hammocks, seeking the ideal hanging location. Most unroll sleeping mats and sleeping bags. Within minutes of the announcement that it is safe to occupy the floor, many already sleep and will sleep literally until we begin our descent into Germany.

I used to wonder at the relish with which the soldiers sought to shape this *point of culminating preparation*. Sure, comfort for the nine-hour flight was important but some of the soldiers almost seemed to look forward to it, as if they literally couldn't wait to get on board and get some sleep. It has occurred to me now why perhaps this is the case as I've examined myself and my own need to secure my floor space and the angst in my own heart until the suitability of floor space could be confirmed.

The point of culminating preparation is the point where they have done absolutely everything they need to do to get to where they need to be, including arriving at the intended point of destination in the best possible condition. Everything else now is beyond their control. They have shaped events to the greatest possible extent down to where and how they sleep on the flight, and now the pressure is off and they can finally relax and let the system work. Preparation has been arduous and difficult, often the culmination of a long process, a treacherous path. These last few steps provide the icing on the proverbial cake, the reward for the efforts being a restful flight. It is no wonder they sleep so soundly.

You can observe the immense satisfaction in the soldier as he rolls over, proudly pops his Ambien, and drifts off into sleep. It's almost as if this is the exclamation point on preparation. This is the last barrier, the final hurdle between them and what they have been called to do. They have done their part to get there, their fate now firmly in the hands of others, those who would deliver them to the battle.

I've got about 8 hours until Germany and my Benadryl is kicking in. This bedroll sure feels nearly as glorious as my bed back home, but not quite. When I wake, it'll be nearly time to punch the clock.

10. The Utility of War

On Being Watched

It just never occurs to me exactly how much people watch one another. As a people-watcher of the highest order, I find great joy in airline travel. Now, I hate to fly commercial, the flight portion of it anyway. Confinement to a typical passenger jet seat feels more than a bit like torture.

I love airports. I love to watch people in airports. I love to sit, drink coffee, and observe the κόσμος, the *cosmos* that is the broiling mass of humanity. Sometimes I wonder if people watch me as much as I watch them as I always assume anonymity. Why on earth would anyone watch me? Surely I'm not the only people watcher out there.

As a commander, I really don't think I comprehended how much my soldiers watched me. I distinctly recall watching all of my commanders, studying them, listening to them, breaking apart and analyzing everything they said. Surely my soldiers and officers did the same with regard to me.

As much as I watched myself and observed myself as I re-engaged with the enemy following my conversion, I'm positive it didn't hold a candle to the amount of time my wife spent observing me. At some point, I came to understand that she was watching me intently, studying my every move. It came to a head at the mall.

Due to the size of our family and the sheer busyness of life, Ami and I have but a few chances here and there to be alone. I find tremendous pleasure in just strolling around any particular place with my wife, having no destination, reveling in her presence. This provides great rest to my soul not to mention that my wife is a great people-watcher as well.

This one particular day between deployments, we made the 40-minute drive to the mall and were strolling around casually, chatting idly, relaxing a bit when she confronted me. I don't recall the specific details of the conversation, only that she wanted to know what was going on with me?! I wasn't talking to her. I'd come to Christ and been to combat several times and had not confided in her about anything. I just continued on about my business as if nothing was going on! Well, what was going on?!

Surely I struggled internally. Surely the demons of conflict and combat tore at my soul. Surely my mind struggled to reconcile the new reality of my faith with the horrors of war. She poured it on and I remember standing there speechless for a few minutes, stunned. I suspect that I am like many men in that I find myself sometimes completely out of tune with my wife's thought patterns and amazingly, even after she has repeatedly explained things to me.

This was the case now. I likely pondered deep truths about what we should get for lunch. "Could I justify both pizza and Chinese food?" as my wife labored in her heart, worried about the condition of my soul. Had war damaged me?

After a minute, I collected myself and explained to her that I hadn't been blowing her off in sharing how I feel it's just that there was nothing there. I felt…absolutely nothing. It wasn't a dry sort of emptiness, an unhealthy emptiness, or a numbness or callousness.

I felt remorse that men had died by my hand and by the actions of the team I had been a part of, and I prayed fervently that I had been an instrument of deliverance rather than an instrument of justice but that God would get the glory either way. I felt regret that the war had kept me from my family for so many months and years. I felt fatigue, knowing that my near future included yet another rotation and that I saw no end in sight. I felt satisfaction that I had performed decently in the crucible and even more to have been a part of a great combat team, one the likes of which truly had never been seen before. I felt that all of that was remarkably normal and aside from that I felt…nothing.

Could I call it peace? Perhaps, more like contentment tempered with resolve.

That They Might Know War

As we've examined the concept that God <u>tests</u> through warfare, let us now shift to a different though not unrelated idea, that God <u>teaches</u> through war. Let us return to our text.

> *Now these are the nations that the LORD left, <u>to test</u> Israel by them, that is, all in Israel who had not experienced all the wars in Canaan. It was only in order that the generations of people of Israel might know war, <u>to teach</u> war to those who had not known it before.*
> *Judges 3:1-2*

Recall who left Canaan in the land. Joshua and Israel failed in their perseverance and left Canaan. God decreed and they failed in following that decree. In the spirit of concurrence, as clearly stated in verse 1 above, the LORD left Canaan in the land. The will of Israel served the will of God and was enslaved to it. Now, we don't know exactly how this is the case, only that this is the case.

God clearly states the purposes for the concurrence of this action. He left them to test Israel and more specifically, to test all who had not fought, those "who had not experienced all the wars in Canaan." (v.1) In His sovereignty, He tests. Warfare is difficult. Confronting evil requires strength and courage. The weak and timid stand by and allow evil to thrive. Disobedience in this regard produces intense, severe, and lasting consequences. God allowed Canaan to remain, despite the fact that it clearly violated His *revealed* will. He allowed Canaan to remain to test Israel and "to <u>teach</u> war to those who had not known it before." (v.2b) He desired that they (Israel) "might know war."

The startling implication is that knowing how to fight, how to conduct battle, how to execute war is a useful skill, beneficial at some level.

I have a nephew, my brother's son, who at one point contemplated a military career and so he called upon me for advice. His parents, concerned about his future, recommended that he join the Air Force and learn a skill that would translate to future employment in the private sector. Instead, he toyed with the idea of joining the Army and seeking assignment to the infantry, maybe even the Rangers. The call to fight weighed heavy on his heart so he sought consultation with me.

I listened to him spell out the pro's and con's of each course of action before telling him that he should do what he wants, despite what his parents suggested, and if he wants to join the Army and close with and destroy the enemy in battle, then that's what he ought to do.

"But what about learning a skill and all that," he asked.

"What greater skill is there than killing people?" I reassured him, slightly in jest but slightly not. I'm not sure his mother, my sister-in-law, appreciated my candor.

As we've seen, the God of the Bible is a God of War as much as He is a God of love, that He is a fierce Warrior. So it is that David, in preparing for combat, turns to the LORD,

Blessed be the LORD, my rock, who trains my hands for war,
and my fingers for battle;
Psalm 144:1-2

Solomon writes,

For everything there is a season, and a time for every matter
under heaven...a time to love, and a time to hate; a time for war,
and a time for peace.
Ecclesiastes 3:1,8

War has its time as much as any other human activity, perfectly legitimate in its existence, necessary for testing and teaching. The existence of the enemy underwrites this necessity.

On the Incorporation of Evil

They gather daily. The enemies of righteousness do not rest. Day and night, the enemies of God plot and scheme, probing for weakness, searching for conditions to exploit.

The wicked watches for the righteous and seeks to put him to
death.
Psalm 37:32

Satan himself, "prowls around like a roaring lion, seeking someone to devour." (1 Peter 5:8) He consistently seeks to enslave the wicked nature of men, to incorporate evil, to organize.

Consider violence for a minute. Humanity frequently regards violence, particularly violence against another human, as the pinnacle of evil, the ultimate expression of badness. Various aspects of violence may modify the perceived severity. Most understand violence in degrees. How many offenders were there? How many victims were there? Were the victims vulnerable in comparison to the offender? What exactly did they do?

Thus most would view a shove as less serious than a beating, a beating as less severe than a stabbing, or a killing. It is more evil to kill several than to kill one, to kill a child than a man. Various other considerations further stratify the act of violence as to exactly how evil it is.

In what manner did they commit the act? Did the victim suffer? Did they kill them quickly? All would acknowledge the badness of rape and the evil of murder. What about thousands of rapes and murders executed systematically, organized under authority, with corporate intent?

A Case Study

> Author's note: the following section contains true but highly graphic accounts of wartime atrocity. I felt it necessary to include them to appropriately illustrate the dangers of the incorporation of evil. However, one could easily skip this section and lose nothing.

As the Japanese Imperial Army marched on Nanking, China in December 1937, one simple enabler facilitated their rapid advance. According to an embedded journalist, "The reason that the (10th Army) is advancing to Nanking quite rapidly is due to the tacit consent among the officers and men that they could loot and rape as they wish."[29]

In August 1937, Shanghai had fallen after a bloody siege involving over a million soldiers and in November, the Japanese Imperial Army maneuvered toward Nanking, the Chinese capital. As Shanghai had been a devastating and bloody battle, Chiang Kai-shek knew that Nanking would fall and he could not risk the remainder of his forces.

On the advice of his German advisors, he sought to preserve his forces and draw Japan into a protracted land battle in the interior. This would leverage the vast territory of China to sap Japanese will and capability akin to Russia's attrition of Hitler's forces a number of years later. Therefore he left Nanking largely unprotected in the face of the impending Japanese onslaught.[30]

Extensive Japanese efforts at Chinese dehumanization would prove quite effective. This, the extensive losses during the bloody siege of Shanghai, and a 400 kilometer forced march effectively set the conditions for the 'Rape of Nanking', whereby the Japanese Army systematically

---

[29] Joseph Cummins, *The World's Bloodiest History*, (Beverly: Fair Winds Press, 2009), 149.

[30] Higashinakano Shudo, Kobayashi Susumu and Fukunaga Shainjiro, "Analyzing the 'Photographic Evidence' of the Nanking Massacre (originally published as Nankin Jiken: "Shokoshashin" wo Kenshosuru)," (Tokyo, Japan: Soshisha, 2005) 15, accessed March 11, 2016, http://www.sdh-fact.com/CL02_1/26_S4.pdf.

slaughtered as many as 300,000 unarmed combatants and civilians following the fall of the city on December 13. Though horrific, the generalized synopsis effectively sanitizes the incident. For those unexposed to mass atrocity, the scope of evil at Nanking, the magnitude of its implications, proves difficult to fathom. Perhaps a few details will illuminate our perspective.

The diary of John Rabe, a German diplomat, and a 16mm film documentary by the American missionary John Magee coupled with various eyewitness accounts chronicle the six weeks of the incident. The tidal wave of horror began as Japan closed on the city. The incident was appropriately labeled 'The Rape of Nanking'.

The International Military Tribunal for the Far East, war crimes trials convened following the war, estimated that Japanese soldiers raped 20,000 Chinese women.[31] Soldiers systematically went door to door, searching for women to gang rape and kill, often mutilate. Various accounts report that they frequently punctuated a gang rape by penetrating the victim's vagina with various items including bamboo, bayonets, broken bottles, whatever happened to be at hand.[32] Following rape and mutilation, they would often bludgeon the victims to death as their comrades looked on, cheering.

The American surgeon Robert Wilson chronicled the atrocities in a letter to his family. In one episode, he records that,

> Let me recount some instances occurring in the last two days. Last night the house of one of the Chinese staff members of the university was broken into and two of the women, his relatives, were raped. Two girls, about 16, were raped to death in one of the refugee camps In the University Middle School where there are 8,000 people the Japs came in ten times last night, over the wall, stole food, clothing, and raped until they were satisfied. They bayoneted one little boy of eight who (had) five bayonet wounds including one that penetrated his stomach, a portion of omentin (sic) was outside the abdomen. I think he will live.[33]

---

[31] Judgment International Military Tribunal for the Far East, para. 2, page 1102, accessed March 11, 2016, http://www.ibiblio.org/hyperwar/ PTO/IMTFE/IMTFE-8.html.

[32] Gao Xingzu, Wu Shimin, Hu Yungong, Cha Ruizhen, *Japanese Imperialism and the Massacre in Nanjing*, Chapter X, Widespread Incidents of Rape, accessed March 11, 2016, http://museums.cnd.org/njmassacre/njm-tran/njm-ch10.htm on 11.

[33] Zhang, Kaiyuan. *Eyewitness to Massacre: American Missionaries Bear Witness to*

Consider Wilson's phrasing. Young girls were 'raped to death.' I cannot imagine the brutality, the horror and agony of being raped to death, literally penetrated until you die. I think of my 17-year-old daughter and an unimaginable fate like being raped to death. I struggle to fathom such evil. John Rabe, the German diplomat, recorded,

> Two Japanese soldiers have climbed over the garden wall and are about to break into our house. When I appear they give the excuse that they saw two Chinese soldiers climb over the wall. When I show them my party badge, they return the same way. In one of the houses in the narrow street behind my garden wall, a woman was raped, and then wounded in the neck with a bayonet. I managed to get an ambulance so we can take her to Kulou Hospital ... Last night up to 1,000 women and girls are said to have been raped, about 100 girls at Ginling College. . . alone. You hear nothing but rape. If husbands or brothers intervene, they're shot. What you hear and see on all sides is the brutality and bestiality of the Japanese soldiers.[34]

A scene from John Magee's film records the following incident,

> On December 13, about 30 soldiers came to a Chinese house at #5 Hsing Lu Koo in the southeastern part of Nanking, and demanded entrance. The door was open by the landlord, a Mohammedan named Ha. They killed him immediately with a revolver and also Mrs. Ha, who knelt before them after Ha's death, begging them not to kill anyone else. Mrs. Ha asked them why they killed her husband and they shot her. Mrs. Hsia was dragged out from under a table in the guest hall where she had tried to hide with her 1-year-old baby. After being stripped and raped by one or more men, she was bayoneted in the chest, and then had a bottle thrust into her vagina. The baby was killed with a bayonet. Some soldiers then went to the next room, where Mrs. Hsia's parents, aged 76 and 74, and her two daughters aged 16 and 14 (were). They were about to rape the girls when the grandmother tried to protect them. The soldiers killed her with a revolver. The grandfather grasped the body of his wife and was

---

Japanese Atrocities in Nanjing. (M. E. Sharpe, 2001).

[34] John E. Woods, *The Good Man of Nanking, the Diaries of John Rabe.* (1998), 77.

killed. The two girls were then stripped, the elder being raped by 2-3 men, and the younger by 3. The older girl was stabbed afterwards and a cane was rammed in her vagina. The younger girl was bayoneted also but was spared the horrible treatment that had been meted out to her sister and mother. The soldiers then bayoneted another sister of between 7-8, who was also in the room. The last murders in the house were of Ha's two children, aged 4 and 2 respectively. The older was bayoneted and the younger split down through the head with a sword.[35]

Another survivor wrote,

The seventh and last person in the first row was a pregnant woman. The soldier thought he might as well rape her before killing her, so he pulled her out of the group to a spot about ten meters away. As he was trying to rape her, the woman resisted fiercely ... The soldier abruptly stabbed her in the belly with a bayonet. She gave a final scream as her intestines spilled out. Then the soldier stabbed the fetus, with its umbilical cord clearly visible, and tossed it aside.[36]

The soldiers divided thousands of women like cattle and organized them into grades in "pleasure chambers" to service as many Japanese soldiers as they could accommodate. Many women committed suicide or went on hunger strikes until they died. Numerous Japanese troops actually photographed themselves immediately following a rape, forcing the women to reveal their genitals while they smiled and posed, often accompanied by a friend or two.[37]

Account after account documents the systematic, *organized* brutality. Mass executions of unarmed combatants and civilians alike continued unabated for weeks. The Japanese spared neither the young nor the old, acting with equal part impunity and aplomb. Only the establishment of a provincial government in February 1938 effectively ended the massacre, at least in Nanking, as it continued elsewhere throughout China as the Imperial Army marched on.

---

[35] Ibid., 281.

[36] Celia Yang, "The Memorial Hall for the Victims of the Nanjing Massacre: Rhetoric in the Face of Tragedy" (2006), 310, accessed March 11, 2016, http://bootheprize.stanford.edu/0506/PWR-Yang.pdf.

[37] Gao Xingzu, Wu Shimin, Hu Yungong, Cha Ruizhen.

About the Organization of Evil

In conducting research for this work, the documentation for Nanking proved as troubling as it was extensive. Some have questioned the death count and extent of the atrocities. Certainly a manner of ambiguity exists as records from such a turbulent time are not meticulously kept. Much documentation was lost or destroyed. Unsurprisingly, Japanese nationalists vociferously and repeatedly deny the claims of atrocity. Despite debate concerning the extent, something happened, a great and terrible something that leaves the student of history with a singular, troubling question. How?

How could something like this happen?

Numerous photographs accompany the commentary concerning Nanking. Pictures of piles of severed heads, beheadings in action, firing squads, beatings, piles of bodies, executed babies, raped women and many more unspeakable images litter the extensive accounts. One particular image haunts my soul, that of a Japanese soldier bayoneting a young boy, maybe two years old, in mid-air. It appears as if someone tossed the child into the air for the soldier to skewer on his bayonet. Again, the question of, "How?" confronts us.

How could any man do this? Does this man have a soul?

With individual acts of evil, the shock wears off. People commit shameless acts of evil all the time, every day. In their prevalence, they just don't surprise. We may struggle a bit, ask some questions, and wonder how? Ultimately, we chalk it up to the hardness of our world. Reckoning the magnitude of a collective fall such as at Nanking confronts us with a nearly impossible quandary.

From the Fall, sin corrupts exhaustively, from the hearts of men to every aspect of creation, tainted and twisted into a mockery, a shadow of God's original intent. From our exposition of Genesis we examined the progression of the curse into the first act of violence, that is the first murder. We noted the troubling generalization of the Fall and its universal applicability: that all men stand corrupted. All men serve a sinful, wicked heart. Though it may resonate and manifest itself differently in different men, the capacity exists uniformly. This understanding demands two unsettling conclusions:

1. All men possess the capacity to rape and kill and at some level, the desire.

2. Organization and collaboration amplifies the effects of raping and killing.

Nanking confronts our sense of righteousness and wickedness. The image of the soldier bayoneting the baby demands numerous unanswerable queries. Obviously, this man is a monster but consider that this man was someone's son. He had a mother, maybe a wife. Someone loved him, trusted him. What about his accomplice who had thrown the child into the air, also obviously a monster? He had a mother, maybe children of his own. Someone likely loved this man and trusted him. What about the by-standing soldiers? The photographer? Were they all monsters? What about the tens of thousands of Japanese soldiers who conducted the Rape of Nanking? Were they all evil monsters?

Why didn't anyone say anything? Why didn't anyone do anything? Nanking insists upon some insight into normalization, group dynamics, and legitimization. I don't intend a psychological assessment of the corporate aspects of wickedness as I do not possess the qualifications. Clearly, evil incorporated and organized is evil amplified. As we assess all things from a proper spiritual perspective, Satan clearly relishes the organization of evil and seeks this ahead of all other things.

The rape of a child serves satanic purposes. A collection of men raping children serves his purposes in a much greater way, to his delight.

Even the casual student understands the rampant historical proliferation of organized evil. Hitler and the Nazis sought to exterminate the Jews, slaughtering more than 6 million. At least 50 million perished under Stalin. Pol Pot and the Khmer Rouge slew millions in the killing fields of Cambodia. Executions, starvation, and other atrocities took the lives of at least 3 million during Mao's Cultural Revolution.

It's not as if these tyrants cut throats themselves; blood never directly stained their hands. They had, as their accomplices, millions of willing and nameless men—fathers, sons, and husbands. Millions of ordinary men surrendered their humanity and goodness on behalf of a collective evil. The collectivization of evil provides Satan with a perfect tool to inflict the absolutely maximum amount of misery upon humanity.

Back to Nanking. As Nanking provides a stark and shocking view into the reality of corporate evil it proves useful in another way. Nanking forces us to examine our response.

What would we do?

On the Necessity of War…a Godly Lesson

The existence of evil demands a response. As evil cannot be appeased, it must be dealt with. One is thus forced to submit, condone, or resist.

Collective evil necessitates an even greater urgency and measure of response.

The presence of evil in this world necessitates resistance, fighting this evil, overcoming. Both individual and collective evil require a collective response. Our moral clarity stems from the knowledge of that which is good and that it comes only from God in heaven.

Wicked men seek to destroy that which is good, that which is righteous, that which is godly. When we understand and account for this simple truth, we quickly realize the necessity of fighting, the value and utility of war.

When confronted with individual evil, the ability to fight proves useful. In the face of a collective evil seeking to impose its will upon the multitude, martial skill becomes infinitely more valuable and decisive.

Some will maintain, "War is not the answer." Those called to fight would rather it not be the answer. No man desires peace more than the soldier, but peace is not the reality of this fallen world.

Consider the major wars fought by the United States. Yes, this government is far from perfect and is governed by imperfect people. Consider a macro view, ignoring for a moment the individual sin of men. The first major war, the Civil War, ended the tyranny of slavery. This nation slaughtered itself to end slavery. The corporate shedding of Nazi and Japanese blood effectively saved the Jewish race and ended the Rape of China. In Vietnam, the United States opposed the evil horrors of communism. In the Gulf War, the US confronted the brutal dictatorship of Saddam Hussein. In the current wars, the United States and its Allies confront the militant face of Islam, an intensely diabolical system.

Let's talk pragmatism and sin for a minute. In Section 5, "War and the Nations", I'll examine the issue in more detail, but I'd like to address the burgeoning cynic at this time and at a minimum allay concerns for the later discussion. I'll concede that the above wars were executed by imperfect administrations with impure motives and intentions at all levels. Does this render them illegitimate?

Consider the Church and the well-worn argument critics use to denounce the body of Christ. It's full of hypocrites. I'll concede that the Church is indeed full of hypocrites; it's full of liars, swindlers, adulterers and all manner of sinners. It's full of people, God's people. Despite the inherent sin, the Church is the hands and feet of Christ. God has great love for the Church and uses the Church to accomplish His will. *Iglesia*, the Greek word for Church, literally means *called out ones*, and God is seeking to call out others to join the Church in spite of its imperfections or perhaps, because of its imperfections.

We've seen that God enslaves all things to serve His will including imperfectly justified or executed wars, of which are all of them. Even obvious confrontations against evil such as the total war of World War Two are riddled with tragedy, sin, and corruption at many levels. Does this negate the overarching good of defeating the Nazis and fascism? Should we yield to the collective evil merely on account of the fallibility of the men and the resulting institutional fallibility on a particular side?

God calls the believer to confront evil in three places: in his own heart, in person, and corporately in submission to governments. I'll leave the discussion of the necessary value judgments to later, but for now declare that God uses our intentions, even the impure intentions of well-meaning men, to accomplish His will and in these cases, to defeat evil through the practice of war.

I would argue that few things are more useful than war, knowing how to fight. God's word teaches us this lesson, and the presence of evil that cannot be appeased well affirms it. A personal confrontation might just settle the matter.

Evil in Person

I've seen the face of evil. We were executing a vehicle takedown in the broad and expansive western Iraqi desert. Sometimes we allowed the enemy to vote, sometimes we didn't, depending upon a number of factors. Sometimes the enemy voted before even given the chance.

On this particular assault, as chalk one closed on the target vehicle, the pilots and ground commander seeking final verification, the crew chief preparing to issue the warning shots, a single fighter emerged from the left rear window spraying from his AK-47. He would not yield without a fight. At the command of the chalk leader, the crew chief delivered, at a rate of several thousand rounds per minute, a volley of 7.62 fire that tore into the man and vehicle. It careened to a stop as chalk two, in a cloud of dust, infilled the assaulters.

The first radio call came, "objective secure," followed after a minute by, "feathers and chargers down." This man had hidden his wife and children under a blanket in the backseat. The 7.62 rounds accomplished their purpose on him and tragically, his family. This man willingly ended his own life and that of his family that he might get a shot or two at us. No one spoke a word on the twenty-minute flight back to our base.

I don't consider myself infallible. I still see, even in salvation, where sin governs my flesh. However, I could never fathom this level of evil, putting your wife and children in the line of fire that was meant for you.

I remember a beach trip a number of years ago and my young son, maybe two at the time, running back and forth at the retreating and advancing surf, squawking in delight. At one point, he mistimed his run and got caught in the advancing surf, losing his balance. As he stumbled, he choked out a "Daddy" reaching blindly upward, not even aware of my location. As I was standing right with him, I caught his hands and boosted him out of the surf and into my arms before his butt even hit the sand. My son trusted his very existence to me without even thinking about it. How could any man betray that trust and love in such a manner?

I prayed that the next time our bullets might find their mark much sooner.

11. The Mind of the Warrior

A Man Afflicted

No man should have to spend more than a day in a hell-hole like Ramadi, Iraq. From 2004 to 2007, SSG Stephen Conway languished there for the better part of 26 months wondering if he would ever see his wife again.

Articulating what direct combat does to a man presents a daunting task. All things shape us in some way. In many ways, we are the summation of our experiences, from the poignant but trivial to the major events that bear upon our minds for the remainder of our existence, coloring our conscience, undergirding our thoughts.

Combat is no exception. No man can taste such savagery and walk away unscathed. No man other than the psychotic could pick up a body part from the street—a hand perhaps that moments before had been attached to a living, breathing human being, knowing that perhaps you had been the one whose actions had led to its detachment—and not be affected in some way. That's just not realistic. The question becomes, "What do you do with something like that?"

It's a question that Stephen Conway proffered several times as we chatted about his experiences over several cups of Starbucks coffee. Conway, focused and intense, candidly bared his soul for nearly four hours. His voice rarely departed from a steady, conversational tone though I wondered what the middle-aged man working at the next table, definitely within earshot despite Starbucks' best efforts to drown out all conversation, must have thought as our discussion ranged from the latest concerning football to matter-of-fact descriptions concerning the taking of human life punctuated by Conway's frequent need to stand and either remove or don his jacket. Here was a man afflicted by the thing that is war.

At one point several years ago, following his second tour to Ramadi, the Army assigned him to "light" duty as a recruiter. Inevitably the anger, the frustration, the pent-up stress, the unexplainable tension began to boil over.

As his unit occupied a new office, he secured a telephone for his desk from the box that contained all the phones, only they were marked. A co-worker informed Conway that he had grabbed the wrong phone.

"It doesn't matter."

"See it's got my sticker. You got my phone."

Conway literally snatched the phone off the desk and tore the chords from the wall, flinging it at his surprised co-worker. Things at home likewise deteriorated. He drifted from his wife, aloof and callous. In frustration, he ripped his young son's shoes off his very feet. He began to drink, hitting the club with the always available party crowd.

Finally, a concerned boss intervened and, as Conway acknowledged, likely saved his life.

Pursuit of Knowledge

Despite popular misconceptions, Christianity is a thinking man's religion. It's always been that way though I didn't come to that realization until several years after my conversion. Prior to conversion, my unregenerate mind bristled at the notion. An encounter in a dark field one night in the Republic of Korea highlighted the extent of my ignorance.

As a young lieutenant, I was serving as the Air Mission Commander for a company level air assault of some 2nd Infantry Division soldiers, maybe a four aircraft mission. We arrived at the pickup zone (PZ) early and had about two hours until sunset, so we did what soldiers do best, we smoked and joked. Standing around this darkening field in Korea, the conversation eventually strayed to religion and somehow to that most cantankerous of issues, creation versus evolution.

I sat and listened in disbelief as a young warrant officer explained his biblical creationist perspective and I do believe this was the first time I'd ever heard such rubbish. Growing up, my church never taught anything like this that I recall. My public education had firmly entrenched my mind in the evolution camp to the point that I didn't believe that people actually existed who denied evolution. I would have categorized them as similar to those who believed that the earth was flat, a long extinct people group. I believed evolution was as indisputable as the roundness of the Earth. I couldn't believe anyone would believe differently in the face of indisputable facts.

I stood there and listened to this young warrant officer, this hick as I thought of him then—he had a strong North Carolina accent—speak about creation and six days instead of evolution, and thousands of years instead of millions and billions of years, and I became incensed. I raged in my soul. His sheer ignorance in the face of obvious fact infuriated me.

How could he believe this?

How could he be such an idiot?

How could he see the obvious facts and draw these erroneous conclusions?

I don't recall the exact nature of my response but I persecuted him. I lambasted him. I questioned his intellect. I did everything but physically assault him, and I was supposed to lead this man.

Needless to say, at the secular military level, this incident could definitely serve as an example of how NOT to effectively lead your soldiers. However, at the level of spiritual reality, another issue governed the engagement, the pursuit of truth.

The Bible encourages the pursuit of truth. God has no secrets, nothing to hide and thus Paul writes,

> *Do not be conformed to this world, but be transformed by the renewal of your mind, that by testing you may discern what is the will of God, what is good and acceptable and perfect.*
> *Romans 12:2*

Paul calls the believer to forsake the conformity of this world and pursue truth, being in fact, transformed in the renewal of the mind. The pursuit of God and the study of His word produces transformation. Instead of seeking truth for truth's sake, the believer seeks truth that he might yet be transformed by it, continually conformed into the image of Christ, and to what end?

Paul tells us that the more we are transformed, the more able we become to discern the good, acceptable, and perfect will of God. (Romans 12:2) Pursuit of God and study of His word transforms our minds into His likeness offering us insight into the mind of Christ, enabling us to see, discern, and understand His will, what He would have us do.

All of the great fathers of science saw their pursuit not as seeking to undermine God and the Bible, rather as further revelation of the beauties and wonders of the Creator and His creation. Kepler, Galileo, Newton,

Pascal, Boyle, Faraday, Maxwell, and Mendel—all professed Christ *and* pursued truth and knowledge with passion and conviction. Nowhere does the Bible call us to blindly accept anything unreasonable.

Faith governs Christianity, driving our salvific belief in Jesus Christ and His atoning work on the cross at Calvary. Outside of the seeming folly of salvation itself, something which the unregenerate can never grasp, logic and reason govern the Bible and why wouldn't it? It stands to reason that revelation of a God of logic, order, and reason would find a base in logic, order, and reason.

I came to Christ not as a student of God's word but as a student of science and reason. I prided myself on my logic. I come from a line of engineers, a collection of math and science type individuals. My father served over forty years in the nuclear industry and actually holds two patents on machinations of a boiling water reactor that I can scarcely pronounce. He is truly one of the most intelligent men I've met and my brother followed him into the field where he is the second-in-command of the huge company my father worked for. I vividly remember doing Calculus with my father at the kitchen table in the eighth grade…for fun.

God drew me to him and granted me faith in spite of my calculated and analytical tendencies. I went to Christ in spite of my firm grasp of reason and rationality. As I poured myself into my new faith, as I pursued God and His word, at some point I came to understand the compatibility of academic pursuit, of the attainment of knowledge, with the pursuit of God. God is the author of knowledge. The minds of men function, actually grasping that which is reasonable, because that's how God fashioned us, in His very image and likeness.

A Cracked Foundation

One particular night rocked the core of Stephen Conway's faith. Though his first year-long tour to Ramadi had been a nightmare, it was the second that broached the hellish. That night mid-way through his second tour epitomized the helpless frustration known and shared by countless foot soldiers from millennia past. That night inaugurated his descent into affliction.

Ramadi, the capital of Al Anbar province, lies about 60 miles due west of Baghdad in a strategically significant location along the Euphrates River. A city of nearly 500,000 in the heart of the Sunni Triangle, after Fallujah, it became one of the most hotly contested and dangerous cities in all of Iraq.

As the Sunni insurgency raged, Ramadi and Fallujah fell to nearly complete Al Qaida domination. The Sunni Awakening eventually facilitated the 'liberation' of Ramadi as the tribes became increasingly weary of Al Qaida's atrocity and turned against them.

I had rarely seen such devastation as from Fallujah to Ramadi. Looking like something from the Apocalypse, every street was filled with rubble, garbage, and filth. Each building bore the scars of years of incessant warfare as the insurgency raged from 2004 to 2007. The enemy lurked in every shadow, around every corner, behind every door.

From Shark Base along the Euphrates, our task force launched repeated assaults into downtown Ramadi and the endless surrounding palm groves and date farms killing and capturing thousands of enemy fighters. Following every raid we'd recover the ground force to the relative safety of Shark Base which had the best chow in central Iraq. We'd then make the 30-minute flight to the gigantic Marine base at Al Asad in the middle of the western Iraqi desert.

Conway and his men had no such sanctuary. As a squad leader in a leg infantry battalion, Conway's charge included the survival of his young soldiers and the accomplishment of the mission and really, making sense of it all. As the blood of the dead and maimed, both enemy and friendly, soaked the streets one could not help but question the utility of it all.

During his first tour in Ramadi, Conway's company rarely ventured into the city proper. When they did, he acknowledged that they'd always ensure the enemy knew they were coming so that they could stand by. They never ventured very far into the city and usually just transited to elsewhere. An uneasy truce kind of governed the relationship. It was a we-know-you're-there-and-you-know-we're-here-but-we-don't-really-want-any-trouble-so-if-you'll-just-let-us-go-about-our-business-we'll-let-you-go-about-yours kind of a situation.

Pragmatics often govern the actions of the soldier on the street despite the intentions of higher command and for several months this condition persisted. There were frequent engagements, sniper fire, and the most dreaded of weapons, the IED. Still, the major engagement eluded them, much as they'd have it.

Conway's second tour proved different. The first six months were unlike anything he'd ever experienced or would ever experience as the pace of combat pushed them all to the brink of insanity. Headquarters was determined to 'clear' the city in support of General Petraeus' newly developed and implemented strategy of "Clear, Hold, Build".

Every day was a constant fight as Conway's battalion pushed into the city from the east and the north. The pace and intensity, the persistence shocked even the most battle-hardened soldier. As other coalition forces pressed into the city from the west, Al Qaida dug in between the two advancing forces and fought fiercely for their very existence. A major engagement was brewing.

On that night, third platoon would lead the assault from an open hotel and probe deeper into the city. Conway's squad would take point. Eerie quiet greeted them as they moved in on foot with the objective of occupying an over-watch position of an AQ financier's home whereby they would call in an airstrike and demolish the home with an LGB, a laser guided bomb. Despite their objective, the commander decided to clear every building adjacent their route, leaving nothing to chance. Though understandable, this slowed their movement to the objective allowing the enemy time to muster.

Building after building they moved in and cleared and again, not a single person, nothing. They were all empty. A chill went up SSG Conway's spine. ISR reported a battle position one alley over so the lieutenant sent Conway's squad to reconnoiter. Nothing. Conway marshaled his men near a courtyard when it happened.

Here Conway struggled to describe the explosion as the artillery round embedded in the wall detonated taking most of his squad with it. He was thrown backward several feet and knocked unconscious. Four of his men lost legs in an instant, three losing one leg, a fourth losing both. As the dust from the blast subsided, the 13 men of Alpha Squad lay scattered about the street in various states of injury. Body parts littered the site. Conway came to, slowly. He checked himself for injury, struggled to his feet and set about pulling the wounded into the courtyard, applying tourniquets, assisting the one young medic. In shock, one of the wounded soldiers could not stop laughing.

At this point, the inevitable fog of battle (Murphy) reared its head. They needed their vehicles to evacuate the wounded but also needed them for the mission. Eventually, they decided to pile all the wounded into a single HMMWV. They strapped the double-amputee to the hood. "Like something from a movie," Conway confided in me as he strapped the soldier down while reassuring him, "Hold on, hold on, stay with me, you're not gonna die."

After the CASEVAC, SSG Conway stood dazed on the corner, covered in the blood of his friends, ears still ringing from the blast, looking at the handful of soldiers remaining from his squad when he received the chilling order, "Move out."

A Soldier's Experience

SSG Conway's experiences were so typical that they are actually atypical. He married his best friend's sister, the amazing and soon-to-be bestselling author, Hannah Conway. She was his polar opposite, gregarious and outgoing, idealistic, the Homecoming Queen. Raised by a combination of his grandparents and mother as he never knew his father, Conway found himself on a football scholarship to the University of Toledo.

He dropped out after a single semester as the reality of the competitiveness of division one football coupled with a general dissatisfaction with college life depleted his desire. He just didn't want to be there and so he eventually enlisted in the Army. Though 9/11 had already occurred, Conway was not a part of the wave of hyper-patriotic recruits that joined following the attack. He needed a job. The Army assigned him to the infantry.

He and Hannah married and the Army deployed him to Korea as war in Iraq was beginning to seem imminent. One day, his commander informed Conway's unit that they would deploy direct to Iraq from Korea. Excited at the prospect and admittedly naive, he and the others in his unit reveled at the notion of engaging in combat. They wanted to be tested. "Why train if you're not going to use it? Are we who we say we are?" For Conway, this meant he would spend a second consecutive year away from his wife.

During his 10-day R&R prior to deployment, he and Hannah conceived a son who was born during his first tour in Ramadi. The Army did not release him to return for the birth.

Following the first deployment, Conway received orders to the 4th Infantry Division at Fort Carson, Colorado where he met up with Hannah. She had moved the household in his absence. Things were okay at home but they had never lived together and he generally kept to himself. She knew something was wrong but they didn't talk much. He felt a general emptiness exacerbated by the general tension of knowing he would deploy again soon.

At PT one morning, Conway's battalion commander assembled his soldiers and informed them that they had received orders. They would deploy in a few months. A raucous cheer went out from the group, excited to return to combat when the commander dropped the bombshell. We're going to Ramadi. It was like the air had been sucked out of the place. An audible groan came from the formation. They all knew what Ramadi was like. Many like Conway had served there previously. The commander dismissed the disheartened formation. Conway went home to tell Hannah.

After ten months at home, Conway once again found himself patrolling the streets of Ramadi and he even got caught up in the Surge. From 2004 to 2006, as the Iraq War spiraled out of control and the body count grew, calls from the public and political leaders to cut losses and retreat grew increasingly strident. The war seemed unwinnable. However, President Bush dispatched General Petraeus as the commander and surged tens of thousands of additional soldiers to the war. The Surge coupled with the Sunni Awakening, the tribes' rebellion against AQ, turned the tide of the war. For soldiers like Stephen Conway it meant that as the end of their year-long tour approached and it looked like they'd actually make it, they got the word that they'd extend for another three months. Back home, Hannah anguished in their separation.

Conway made it though and the Army assigned him as a recruiter and after several years and a promotion to Staff Sergeant, sent him to the famed 101st Airborne Division for a deployment to the Konar region of Afghanistan. Six months later, a sniper's bullet found his right forearm and a medical evaluation board awaited him on his redeployment. After over a decade of service, more than three years of vicious combat from the streets of Ramadi to the peaks of Afghanistan, after three combat wounds including a Traumatic Brain Injury and a PTSD diagnosis, the Army cut Stephen Conway loose. He had been on the fast-track, on pace to make Sergeant Major, now he was left wondering what to do with the rest of his life.

Chapter 12

12. The Struggle of War

<u>Upon the Demotion of Man</u>

Modern revelers, in turning from God, exalt man. Paul's admonition in Romans chapter 1 reads like a contemporary narrative as he writes that God handed wicked men over to judgment because,

> *they exchanged the truth about God for a lie and worshiped and served the creature rather than the Creator, who is blessed forever! Amen.*
> *Romans 1:25*

Because of creation, all men know that there is a God yet, "although they knew God, they did not honor him as God or give thanks to him." (v.21) God judges men for denying Him, the Father. All men, in the sinfulness of their hearts knowingly forsake God and therefore have no excuse. (v.20b) They worship and serve the created thing rather than the Creator.

In the denial of God and the subsequent exaltation of man, secularists actually accomplish the contrary to which they set out to achieve. Attempts to exalt the creature actually relegate man to little more than an animal enslaved to his base instincts and lusts.

Without God, secularists must rely upon ungoverned forces such as chance and time to account for all of reality, including man and all of the inherent difficulties that his existence generates. They must account for not only his physicality, his behavior, but intangibles such as morality and conscience and even man's own recognition of the metaphysical, no mean feat.

Science classifies anatomically modern humans (AMH) as *homo sapiens sapiens*, a subspecies of *homo sapiens* not to be confused with our ancestral subspecies *homo sapiens idaltu*, the latter of which purportedly died out 160,000 years ago. Of note, *homo sapiens idaltu* exists as a proposed subspecies, not yet verified by the fossil record.

As *homo sapiens* translates "wise man" the species has performed remarkably well. "The ingenuity and adaptability of *Homo sapiens* has led to its becoming, arguably, the most influential species on the planet."[38] You really cannot fabricate statements and thinking such as this. *Homo sapiens* has become the most influential species on the planet! Arguably?!

Though not considered an academic source I thought the Wikipedia article well summarized secular views concerning man. As the article obviously assumes evolution, man was left in competition with other species for dominion over the planet. I was surprised that the article referred to homo sapiens as "arguably" the most influential species on the planet which begs the questions, which other species possesses a similar level of influence? If homo sapiens so desired, he could destroy every other species, including the entire planet, save for maybe insects. Does durability imply influence? Though laughable when articulated in certain ways, this type of thinking governs almost all modern, secular democracies.

As all modern humans fall into the subspecies *homo sapiens sapiens*, we are "wise, wise men", maybe extra-wise men, lending further credence to Paul's general admonishment in Romans 1,

> <u>Claiming to be wise</u>, they became fools, and exchanged the glory of the immortal God for images resembling mortal man and birds and animals and creeping things.
> Romans 1:22,23

In foolishness, man declared himself wise, as seen in our own labeling, *homo sapiens sapiens*. Thus, we denied God and sought to exalt ourselves, exactly as Paul said we would.

For those who seek to exalt man in this manner, instinct must account for behavior. We must rely upon animalistic instinct to explain exactly why men behave in a certain way. Men become slaves to their stomachs and penises. Women become slaves to their stomachs and their need of security. People exist coincidentally, for no altruistic or metaphysical reasons. Therefore, gratification of the base instincts and primitive lusts necessarily motivates, governing all that they do.

---

[38] "Homo sapiens," accessed at https://en.wikipedia.org/wiki/Homo_sapiens on April 12, 2016.

Forced to explain on these terms, I must hereby account for morality and conscience in terms of evolutionary benefit, collective well-being, or other group dynamics. The logic of this explanation quickly deteriorates when transitioning from individual to the collective actions of men.

Consider rape if you will. As a sexual creature, gratification of sexual lust is one of my basic needs. Monogamy is a false societal construct that serves me, the individual man, no actual benefit. I benefit personally, secularly, only by copulating with as many females as feasible, by force if necessary. This provides me with sexual gratification and the obvious benefit of spreading my seed to as many fertile hosts as possible.

Now, clearly rape is harmful to society and any honest sociologist will concur that monogamy, that marriage is beneficial to society. They cannot explain why, just that it is. They must explain why a man would subordinate that which he would gratify for the sake of a greater collective good. Evolutionarily speaking, why do I care about a greater good? Would not the satiation of my base lusts drive my actions? The concept of subordination bears no logic in an evolutionary construct.

If this analysis drives me toward something transcendent, it forces me into a confrontation with a deeper truth, a universal morality, a code that exists as a hint at something more. Have you wondered why virtually every society in existence has declared murder to be wrong and illegal? Thievery? Rape? Apart from the prospect of a Creator, I must invent to circumnavigate these realities.

Impacts of Demotion

Though not immediately obvious, the demotion of man in this manner, even with the intent to elevate, provides a useful foundation for a dizzying array of ideologies and systems. Many of these have been adopted and utilized toward the destruction of the very creature they've purported to exalt, man.

Stalinists, Nazis, and Fascists all denied a Creator as did Pol Pot and Mao. If Hitler's Third Reich possessed a state religion, it was evolution. The Aryan race epitomized the pinnacle of evolutionary man, the apogee of development. The Jews were the opposite. As *lesser* humans, they inhibited man's development, polluted blood lines, and therefore had to be accounted for. From the first volume of *Mein Kampf*,

> The stronger must dominate and not mate with the weaker,

which would signify the sacrifice of its own higher nature. Only the born weakling can look upon this principle as cruel, and if he does so it is merely because he is of a feebler nature and narrower mind; for if such a law did not direct the process of evolution then the higher development of organic life would not be conceivable at all.[39]

and,

If Nature does not wish that weaker individuals should mate with the stronger, she wishes even less that a superior race should intermingle with an inferior one; because in such a case all her efforts, throughout hundreds of thousands of years, to establish an evolutionary higher stage of being, may thus be rendered futile.[40]

Hitler embraced evolutionary thought and expounded it to its fullest. Perhaps no event personified evolution more than the Holocaust.

Evolution is a bankrupt *philosophy* that drives hate, providing substantiation for the elevation of a particular group or race. As man demotes himself to the highest order of evolutionary development, an accidental development as it were, basis now exists for subjugation.

Note the difference between micro-evolution and macro-evolution. Micro-evolution is a phenomenon that has been observed and recorded. This work disputes macro-evolution concerning the creation of genetic information, the advance of a species, and the never-recorded transition from one species to another.

The capacity exists within man, inherent from his sinful nature, to elevate himself to a certain position. The demotion of fellow man accomplishes for him what he would otherwise set out to obtain. Other religious systems seek to accomplish much the same.

Islam declares all who do not believe in and submit to Allah as *Kafir*, a term of subjugation rife with implications. The *Kafir* is less than human and therefore not entitled to the same treatment as those of the *Ummah*, the universal body of Islamic believers.

---

[39] Adolf Hitler, *Mein Kampf*, Chapter XI, "Race and People," accessed April 14, 2016, https://www.stormfront.org/books/mein_kampf/mkv1ch11.html.

[40] Ibid.

A *Kafir* is to be mocked,

So today those who believed are laughing at the disbelievers,[41]

A *Kafir* is to be beheaded,

So when you meet those who disbelieve (in battle), strike (their) necks until, when you have inflicted slaughter upon them, the secure their bonds, and either (confer) favor afterwards or ransom (them) until the war lays down its burdens.[42]

A *Kafir* is to be shunned,

Believers should not take Kafirs as friends in preference to other believers. Those who do this will have none of Allah's protection and Will only have themselves as guards.[43]

A *Kafir* is evil,

And say: Oh my Lord! I seek refuge with You from the suggestions of the evil ones (Kafirs). And I see refuge with you, my Lord, from their presence.[44]

A *Kafir* is cursed,

They (Kafirs) will be cursed, and wherever they are found, they will be seized and murdered. It was Allah's same practice with those who came before them, and you will find no change in Allah's ways.[45]

A *Kafir* is less than human and must be eliminated, including all those who claim to be of the *Ummah* but actually live under the authority of *Kafir* or apostate regimes while refusing *jihad*.

An editorial in *Dabiq*, the online magazine of ISIS, called for the death of all *Kafir* but specifically those who claim Islam but are not even apostate, ones who denounce Islam publicly. They still claim the Prophet but deny him in their hearts. "There is the religion of Allah, which is Islam, and then

---

[41] Q'ran 83:34.

[42] 47:4.

[43] 32:8.

[44] 23:97.

[45] 33:60.

the religion of everything else, which is *Kufr*."⁴⁶ The article goes on to state that, "The ruling of the person who commits *riddah* (apostasy) is that he is killed, unless he repents before he is apprehended."⁴⁷

The devaluation of man inevitably generates stratification. The system, whichever system dominates, establishes a measured value culminating with a metered intrinsic worth. The system always declares some as more valuable than others, and when applied by the most zealous, necessarily dictates the dehumanization of a body, potentially requiring their extermination. It becomes necessary.

Devaluation frequently provides the fuel to the sinful human heart that drives the previously discussed collectivization of evil. Nanking speaks to devaluation. Treblinka and Auschwitz speak to devaluation. My Lai and Al-Kasasbeh speak to devaluation.

Precisely one system, the way of the cross, combats the demotion of man, devaluation, and the inevitable stratification and collectivized evil that has so bloodied man's history. Only in Christ does equality exist.

No Respecter of Persons

Christ stratifies with some nuanced but essential differences. To be sure, the Bible declares that all persons are either of Christ or they are not and that apart from Christ, all men stand equally condemned before God.

As we referenced the salvation of the Gentiles and Peter's address to the centurion Cornelius, he opens by saying,

> *Truly I understand that God shows no partiality,*
> *Acts 10:34*

Believers come from every single walk of life—every race, tribe, tongue and nation. As Paul writes,

> *There is neither Jew nor Greek, there is neither slave nor free,*
> *there is no male and female, for you are all one in Christ Jesus.*
> *Galatians 3:28*

Christ is the great equalizer. All men need a savior and Christ is the only sufficient savior for all men. One's position in Christ never drives the

---

⁴⁶*Dabiq* magazine, Issue 14, page 8.
⁴⁷Ibid., 9.

believer to subjugate those not of Christ.

1. God exhorts the believer to pursue the non-believer in love, to carry the message of the Gospel to them, that they might be saved. (Matthew 28:18-20, Mark 16:15, Luke 24:44-49, John 20:21, Acts 1:8)
2. Believers have no idea who is of the elect, whom God has chosen for salvation. All non-believers are potential brothers and sisters in Christ.
3. As salvation is of the Lord, believers have no reason to boast. (Ephesians 2:9) Christ modeled humility and demands it of his followers. When considering non-believers and their position before God, Paul writes to believers, "such were some of you." (1. Corinthians 6:11) No one is born into Christ or attains salvation based upon any merit of their own.

As the Gospel demands humility and equality, another important aspect defines intrinsic worth. Christianity places great value on even a single human life.

The Imago Dei

The creation account records the unique aspect of humanity. After six days, God had spoken everything into existence *ex nihilo*, from nothing. Then He says,

Let us make man in our image, after our likeness. And let them
have dominion...
Genesis 1:26

God, after creating everything, directed the Creation of man. Unlike every other thing that God created by His spoken word, "the LORD God formed man of dust from the ground and breathed into his nostrils the breath of life, and the man became a living creature." (Genesis 2:7) God formed man and breathed into him the *nĕshamah*, the breath of life, the spirit of life. God created man uniquely, separate from all of creation. He similarly creates woman from the rib of the man. (Genesis 2:18-22)

So God created man in his own image,
in the image of God he created him;
male and female he created them.
Genesis 1:27

God made man in His very image, unlike any other aspect of creation, the *Imago Dei*. Man is the crown jewel in God's creation. Even the angels, including Satan and his minions prior to their rebellion and subsequent casting out from heaven, "sang together and…shouted for joy" (Job 38:7) when God "laid the foundation of the earth." (v.4) The angels rejoiced at creation and one can imagine their special rejoicing at the creation of the *Imago Dei*, an honor not even bestowed upon them.

All men are created in the image of God. All men bear the image of God: black men, white men, brown men, Asian men, Nordic men, Muslims, Buddhists, Hindus, Christians, and Atheists. God created them all in His image and as such, man affords unique consideration.

Confusion exists in exactly what it means to be created in the image of God. Clearly it doesn't mean that we look like God as God is spirit (John 4:24) and spirit has neither flesh nor blood. (Luke 24:39)

As the *Imago Dei*, God issued man a special and unique charter, to have dominion over the earth. His first command to the couple was to,

> *Be fruitful and multiply and fill the earth and subdue it, and have dominion…over every living that moves on earth.*
> *Genesis 1:28*

God created man for the explicit purpose of exercising godly dominion over all of creation and spreading His image bearers over the entire Earth through procreation and discipleship. God created man uniquely with a unique purpose.

The psalmist speaks to this uniqueness,

> *what is man that you are mindful of him,*
> *and the son of man that you care for him?*
> *Yet you have made him little lower than God*
> *and crowned him with glory and honor.*
> *You have given him dominion over the works of your hands;*
> *you have put all thing under his feet,*
> *Psalm 8:4-6*

God uniquely considered man and made him even a "little lower than God" above all of creation including the heavenly beings. Now, some translations render this as "little lower than angels (heavenly beings)" though the word *elohiym* is almost always translated *God*. Either way, man occupies a special place in the heart of God and even shares aspects of God's attributes in bearing His image.

We can reason as God does (Is. 1:18), love like God (John 3:16, Ephesians 5:25), even hate as God does. Compassion, mercy, grace, fellowship and friendship all reflect the character of God, His image and likeness.

I had a debate with an atheist friend of mine once. We were in northern Iraq and had a mission go bad. Amazingly no friendlies got hurt, but the leadership rightfully called our judgment into question and placed our entire unit on a temporary stand-down. For several days we had nothing to do but work out, sleep, and hang around. Inevitably, discussion turned to religion.

This friend is one of the most intelligent people I've ever known, a great warrior, but one who takes great joy in undermining the faith of ill-prepared Christians. This particular day, he and I sat debating the finer aspects of evolution versus creation and we got to talking about souls. I made a statement concerning the unique aspects of man and that animals don't have a soul, referring to the eternal aspect of a being's existence.

My friend became incensed at the proposition that his dog did not have a soul. He angrily informed me that his dog loved him and possessed a soul, which in retrospect I find curious that an atheist even conceded the existence of the intangible or metaphysical. I should have told him that only man is created in God's image, that in Genesis 2, when God brings each animal before the man to be named, He establishes authority, that Paul writes,

For not all flesh is the same, but there is one kind for humans, another for animals, another for birds, and another for fish.
1 Corinthians 15:39

I should've told him all of these things but I recall telling him that I loved my dog too, but he ate cat turds. I don't remember how the debate concluded.

After the flood, God warns Noah,

Whoever sheds the blood of man,
by man shall his blood be shed,
for God made man in his own image.
Genesis 9:6

God reiterates His original mandate to Noah to "be fruitful and multiply" and then issues this warning against the shedding of blood. Interestingly, the first visible manifestation of the Fall occurred when Cain shed his brother's blood, and by the time of the Flood sin permeated all of humanity as evidenced by the proliferation of violence and the shedding of blood.

In the Law, the Ten Commandments, we see further indicators of God's view concerning man and His stipulations concerning our treatment of our fellow man. Jesus sums the entirety of the Law into two commands,

1. Love the Lord with all your heart, soul, and mind and,

2. Love your neighbor as yourself.

The first four commandments indeed denote man's relationship with God: you shall have no other gods before me, you shall not make for yourself a carved image, you shall not take the name of the LORD your God in vain, and remember the Sabbath, to keep it holy. (Ex. 20) Love God with all that you have and place nothing before Him.

The remaining six govern men's relationship with one another: honor your father and your mother, you shall not murder, you shall not commit adultery, you shall not steal, you shall not bear false witness, and you shall not covet. (Ex. 20) In other words, love your neighbor as yourself.

Love God and then love your neighbor as yourself. Who then is your neighbor? Is it not the one who resides within your sphere of influence? Is this not describing the intrinsic value God places in man?

Man merits special consideration in life, but also in death, especially in death. When man raises his hand against another, he raises his hand against one whom God knew before He formed him in the womb (Jer. 1:5), one knit together in his mother's womb by the very hand of God. (Psalm 139:13) He may not be a child of God in salvation, but he is a creation of the Most High.

Christianity's elevation of man, in subordination to God, promotes true equality in Christ. Every other way relegates in some manner, ultimately leading to stratification, degradation, and even subjugation. God created man to glorify Him and in salvation, adopted men as children of God. Man is truly the pinnacle of creation. In light of the unique existence of man as the *Imago Dei*, his destruction necessitates careful consideration, special handling.

Questioning His Faith

At some point as he wrestled with this very uncomfortable fact, SSG Conway sought to reconcile. One day in Ramadi, he approached his chaplain and asked, "What does God think about me killing all these people?"

Conway had become a believer like so many others, during a youth retreat in high school. He persisted in the faith immediately following graduation into his adult years and even led a Bible study during basic training. Following his first gunfight in Ramadi, he began to drift, slowly, gradually. The pangs of war, separation from his family, the bloodshed, the pointless nature of it all took a toll. He had no time for religion and it just didn't seem like God was too active in a place like Ramadi anyway.

At one point, he did like so many soldiers and consulted the chaplain. Army chaplains are some of the most amazing men I know, dedicated and faithful to the Lord and to the soldiers of this nation. A good chaplain is an incredible benefit to a unit but like all groups, there are good chaplains and then there are not so good chaplains. Conway remembered this chaplain as a jokester, a light-hearted man who didn't seem to take anything seriously.

Unfortunately, he had no answers for a young SGT Conway and minus the answers he so desperately needed, Conway drifted further into his affliction.

13. Killing the Mind

The ground force had the guy surrounded. It looked like something from an action movie. If someone had not seen the actual event as it occurred, you undoubtedly would have known that something had gone down. An overturned tractor trailer, smoke billowing from a vehicle resting at an obscene angle in a nearby ditch, ten soldiers surrounding a single man with his hands up while another group, brandishing more weaponry than you've seen, conducted crowd and traffic control. Needless to say, southern Kandahar had probably not seen much action like this, as the swarm of black helicopters descended from the scorching daytime sun upon a crowded highway. Things got messy quick.

Me and the Gunny circled overhead in the manner of a hawk, ready to roll in hot at the first hint of resistance, of something going wrong. I always loved the Gunny. He was a great pilot with a great fighting spirit, but an absolutely apathetic officer with little regard for all things Army. He walked a fine line, but he made it work. He loved to fight. I heard him say more than once, "I just want to kill people" and he meant it. I'm sure of this from the bottom of my heart and I'm glad he fought on our side.

"Single MAM (military-aged male) with his hands up," came the call.

"I've got him," as we watched from overhead.

I expanded my gaze to the perimeter, searching for threats, examining vehicles, the forming crowd, anxious to get done and get out.

"Shots fired, man down." I looked back down in time to see the man lying in a heap on the pavement. He had voted and the boys had acquiesced.

In the debrief later that day, I asked the ground commander what had happened. He informed me that the man had displayed 'hostile intent' and left it at that. From 300 feet, it's tough to tell exactly what happened but clearly something led the boys to kill this man and I've never been one to question the man on the ground. Our enemy frequently voted against himself, even when surrounded and outgunned.

This one troubled me a bit. Perhaps it was the split second timing, the fact that no weapon was found. I felt no remorse as I remember he had been a mid-level Talib, maybe a cutout for local leader or perhaps another type of facilitator. This man had aligned himself with the enemy and had blood on his hands. He was due a measure of justice.

Resistance to Killing

Grossman notes that a "lack of enthusiasm for killing the enemy" is "a force that is discernible throughout the history of man."[48] This lack of enthusiasm "causes many soldiers to posture, submit, or flee, rather than fight."[49] The vast majority of soldiers, at least since the advent of gunpowder, when confronted with the enemy have chosen other options rather than actually engaging the enemy with the intent to kill. Few soldiers willfully and actively engage to kill.

S.L.A. Marshall's research from World War Two concluded that, "the average and healthy individual…has such an inner and unrealized resistance toward killing a fellow man that he will not of his own volition take life if it is possible to turn away from that responsibility."[50] A normal human has an innate resistance to taking another human life that is difficult to explain. This resistance forced armies over the years to develop varying forms of operant and classical conditioning to overcome these psychological barriers, but not without significant ramifications.

Many of the psychological issues that soldiers suffer stem from efforts at overcoming this resistance to killing. The impacts are what have never been appropriately assessed.

I saw LTC(R) Grossman speak once and though he doesn't come right out and say it and I can find no direct reference in his work, he hints strongly at a spirituality, even a Judeo-Christian ethic. As the highly secular world of psychology restricts his conclusions, I believe he cannot state openly what he knows to be the truth and instead cloaks his conclusions in psychological vice theological language. In reality, the psychological is directly dependent

---

[48] Dave Grossman, *On Killing*, (New York: Back Bay Books, 2009), 29.
[49] Ibid.
[50] Grossman, 30.

upon the spiritual, enslaved to the metaphysical.

The resistance to killing originates with an untaught and innate understanding that God created all men uniquely in His Image. When one takes life, he actually destroys the same as himself and commits a direct assault upon the Divine. How can there not be consequences to overriding such a powerful impulse as this?

Scarring the Soul

Let's address killing for just a minute. Suppose you have killed one way or another. Perhaps you've pulled the trigger, launched the mortar round, or called in an airstrike. God equipped you with strength and courage, and either through conditioning or consecration you've overcome the natural inclination not to kill and accomplished what was asked of you. I'll assume at this point you are not one of the few who actually relishes killing, but are an ordinary man such as myself.

Through this work, I seek to convey the possibility of righteousness in killing, the possibility of enslaving the act of killing to the perfect and pleasing will of God in the conduct of war.

I pray that we would never become comfortable with the idea of killing, no matter how much we consecrate our hearts, no matter how much strength and courage God imparts to us, no matter how much conditioning has overridden our natural inclinations. A dichotomy exists. A rightly motivated soldier eagerly meets the enemy up front where the fighting is hard. A rightly motivated soldier relishes in a proper gunfight. Battlefield glory lies in closing with and destroying the enemy, the very act of fighting. Yet, the death itself is always tragic, the destruction of the Image of God always grievous.

Thus, I may exalt that my brothers and I stood firm against the evil forces of the enemy and defeated them in battle. Simultaneously, I'll mourn the tragic death of whichever young men we vanquished. No matter the outcome or the reason, these were sons and husbands, fathers even. They were men with hopes and dreams. I may never gloat or glory in the desecration of the pinnacle of God's creation.

I mourn in victory as much as I revel. God demands this of the warrior. Each death scars up the soul just a bit.

The Struggle

Life is a harsh struggle for the vast majority of humanity. Out of the meanness and hardness of the world, humanity scratches out an often-miserable existence plagued by the infirmities of creation tainted by the Curse.

As we discussed in Chapter 4, Paul writes,

For the creation was subjected to futility,
Romans 8:20

and as a result of this futility,

For we know that the whole creation has been groaning together
in the pains of childbirth until now.
Romans 8:22

Creation groans under the weight of sin, afflicted and distorted, in bondage to corruption. As much as the proliferation of sin punishes humanity directly, the perversion of creation afflicts humanity indirectly, via secondary means.

Sin inevitably generates trauma, the tribulation of men.

Merriam-Webster defines trauma as, "a very difficult or unpleasant experience that causes someone to have mental or emotional problems usually for a long time." Several factors define trauma per this definition.

Trauma is an experience that is "difficult" or "unpleasant". The countless conditions that produce trauma start to become clear. Natural disasters produce trauma. Circumstances such as death of a relative or a lost job may generate trauma. People doing things to people, unpleasant and difficult things, generates trauma. Think of rape or violence of some form, certainly combat. Consider the actual taking of human life, which we have well established as an exceedingly difficult thing to do. Killing would qualify as a "difficult" or "unpleasant" experience.

The second condition inherent in this definition is that the experience, as it is difficult or unpleasant, generates "mental or emotional problems". An event must generate problems within the person to be considered traumatic. Thus, an experience that may traumatize one person may be taken entirely in stride by the other. In some ways, trauma is relative, an important consideration when assessing conditioning and stress inoculation.

The last factor to consider merely from the definition is the duration of the problems whereby it persists *"usually* for a long time." The difficult experience generates problems that usually endure, but not always, leaving open the possibility of an acute form of trauma.

Combat and Trauma

In his historic work, *The Red Badge of Courage*, Stephen Crane chronicles the combat initiation of a young Union soldier, Henry Fleming. Henry is a member of the fictional 304th New York regiment marching to battle at Chancellorsville. After several days of maneuvering, the young private receives his baptism by fire as the Union lines turn back the very first rebel charge. However, the short-lived exaltation of the soldiers turns to disbelief.

> "Here they come ag'in! Here they come ag'in!"[51]

The second rebel charge deflates the previously exuberant Union lines. A fierce artillery barrage precedes the assault, confusing and withering the ranks. They lose any will to fight and as such,

> The men groaned. The luster faded from their eyes. Their smudged countenances now expressed a profound dejection. They moved their stiffened bodies slowly, and watched in sullen mood the frantic approach of the enemy. The slaves toiling in the temple of this god began to feel rebellion at his harsh tasks.[52]

Henry, who fought decently enough during the first repelled assault, begins to whither.

> His neck was quivering with nervous weakness and the muscles of his arms felt numb and bloodless. His hands, too, seemed large and awkward as if he was wearing invisible mittens. And there was a great uncertainty about his knee joints.[53]

At one point, the youth throws down his rifle and runs, fleeing certain death in assumed solidarity with his comrades.

---

[51] Stephen Crane, *The Red Badge of Courage*, accessed May 30, 2016, http://www.emcp.com/previews/AccessEditions/ACCESS%20EDITIONS/The%20Red%20Badge%20of%20Courage.pdf, 42.

[52] Crane, 43.

[53] Ibid.

A number of potentially traumatic experiences confront young Henry. Danger is the most obvious of them, then the chaos of battle, and confrontation with the actual existence of men who not only desire to harm him but now actually possess the capacity to do so.

I live with the knowledge that a vast number of men around the globe would desire to do me harm for the mere fact that I am an American and a member of the Army. This causes me no trauma for my current perception is that they do not possess the capacity, either by their limitations or by my own anonymity. It is the realization of the capacity for harm that confronts and provides a potential for trauma. The furious rebel assault confronts Henry in this manner, that his own desire for self-preservation runs contrary to the intent of the advancing horde possessing an opposite desire, the desire to do him harm or even kill him.

From this, the prospect of taking human life and its actualization confronts Henry. As these rebels draw near, intent on doing him harm, he is forced into the quandary of necessarily violating his own internal resistance to killing for the sake of the equally strong internal force that is self-preservation. These two governing principles clash directly and violently, confronting Henry; the clash itself might generate trauma.

Witnessing the trauma of others may prove traumatic. In the midst of the battle, Henry sees men fall to his left and right. He sees his officers babbling incoherently, overcome by the friction of battle. He hears and sees the various reactions of the men around him and all of these things affect him to the point whereby he throws down his arms and flees.

As Henry flees, he eventually joins a column of wounded soldiers making their way to the rear area. He finds himself envious of their condition, that they could be wounded and thus leave the battle with honor. He finds himself wishing that he too could have a little "red badge of courage."[54]

Is there any other occupation whereby one might find oneself wishing for injury if only to avoid a potentially much worse fate, a significantly greater trauma?

Moral Injury—A Special Trauma

Closely related to trauma is the idea of *moral injury*. Moral injury involves, like trauma, an event that is extreme or unprecedented including

---

[54]Crane, 60.

the aftermath of such an event. The distinguishing factor for moral injury is that the actions "transgress deeply held moral beliefs and expectations."[55] A moral injury involves a transgression of one's own moral and ethical expectations. It is rooted in deeply held personal convictions as a function of religious or spiritual beliefs, or culture-based, organizational, and group-based rules about fairness and the value of life.[56]

Per the Nation Center for PTSD, in the context of war, the perceived transgression may stem from the act of killing or harming others, participating in atrocity, or indirect acts such as failing to prevent the immoral acts of others. Receiving orders considered immoral could cause a moral injury. These actions could be either intentional or unintentional.

Moral injury involves a tearing of the fabric of a man's soul in some way. Moral injury elevates the severity of trauma, adding a metaphysical component whereby a man has truly violated the foundation of what he perceives to be righteous. Moral injury generates intense feelings of guilt and shame along with anxiety in worrying about potential consequences.

Perception governs moral injury. Basic misunderstandings of spiritual foundations contribute to moral injury. Thus, a man convinced that killing may never be justified would react differently than a man with no religious convictions convinced of the necessity of killing on occasion and the righteousness of his own cause.

My intent, with this work, is to govern future actions by righting perceptions. If I can equip a man with a correct understanding of matters concerning warfare and killing, then that would theoretically influence his future actions whereby he would conduct war correctly, justly. This would preclude a future moral injury as his actions would square precisely with his preconceptions.

Yet, if you've already fought, the potential exists that you've already received a moral injury. As such, our intentions become reparative, that we may assist the warrior bearing such a wound in reconciling and ultimately, repairing the fissure. In understanding the theology of war, we may address the moral injury retroactively.

---

[55]National Center for PTSD, "Moral Injury in the Context of War," accessed March 16, 2017, https://www.ptsd.va.gov/professional/co-occurring/moral_injury_at_war.asp.

[56]Ibid.

Trauma and History

As armies are in the business of winning at war, they have always noted the unique aspects of trauma generated by combat and the associated and various reactions of soldiers to this trauma. Armies maintain a vested interested in understanding and addressing any issue that might affect the availability of combat power. Soldiers' reactions to the unique stressors of combat have always been a matter of concern, even if for entirely pragmatic reasons.

Civil War era soldiers and before who succumbed to the psychological and emotional toll of conflict were often diagnosed with *nostalgia*. Afflicted soldiers reported trouble sleeping, missing home, a general anxiety, and moderate depression. Seeking a physiological explanation, physicians often diagnosed a condition known as "railway spine", actual physical trauma to the central nervous system from the vibrations of rail travel that generated similar symptoms.[57]

During World War One, physicians named the condition *shell shock* adding panic as a possible symptom. Initially, they believed that physical trauma to the brain from the firing and impact of artillery caused the issues, hence the name. Yet the glut of soldiers who had never been exposed to artillery fire but displayed the same symptoms forced a reconsideration. Doctors prescribed a short period of rest followed by an immediate return to the front for those who suffered.[58]

Combat Stress Reaction (CSR) or *battle fatigue* supplanted shell shock as the diagnosis during World War Two. Long surges in operations yielded increasing CSR diagnoses to the point whereby nearly half of discharges were due to battle fatigue.[59] Physicians still treated CSR with a short rest period along with an accompanying expectation that the soldier would return to battle shortly.

It was research on returning Vietnam veterans that prompted the

---

[57] Matthew J. Friedman "History of PTSD in Veterans: Civil War to DSM-5," accessed May 30, 2016, http://www.ptsd.va.gov/public/PTSD-overview/basics/history-of-ptsd-vets.asp.

[58] Adam Hochschild, *To End All Wars - a Story of Loyalty and Rebellion, 1914-1918,* (Boston, New York: Mariner Books, Houghton, Mifflin Harcourt, 2012), xv, 242, 348 accessed May 30, 2016, http://library.umd.umich.edu/newbooks/2012/may.php.

[59] Friedman.

American Psychiatric Association (APA) in 1980 to standardize Post Traumatic Stress Disorder (PTSD) as a diagnosis in the Diagnostic and Statistical Manual of Mental Disorders (DSM-III). Holocaust survivors and rape victims also contributed to the research, and for the first time researchers established a causal link between wartime trauma and post-military civilian life. Subsequent updates to the DSM annotated the prevalence of PTSD and a change in its categorization from an anxiety disorder to the newer category of Trauma and Stressor Related disorder.[60]

By 2016, millions of American veterans have been diagnosed with PTSD and are receiving treatment in a Veteran's Administration (VA) facility.[61] This testifies to culturally widespread acceptance of the legitimacy of behavioral health diagnoses and 16 years of persistent conflict providing a never-ending supply of stressors.

Examining the Issue

As we possess little control over the stressors, the response becomes the issue or what necessitates the *disorder* label. Varied responses to trauma challenge providers in both diagnosis and treatment. The response truly defines the condition.

The National Institute of Mental Health (NIMH) defines PTSD as,

> a disorder that develops in some people who have seen or lived through a shocking, scary, or dangerous event.[62]

The APA defines it as,

> an anxiety problem that develops in some people after extremely traumatic events, such as combat, crime, an accident or natural disaster,[63]

---

[60] Friedman.

[61] Veterans and PTSD, "Veterans statistics: PTSD, Depression, TBI, Suicide," accessed May 28, 2016, http://www.veteransandptsd.com/PTSD-statistics.html.

[62] National Institute of Mental Health, accessed May 30, 2016, http://www.nimh.nih.gov/health/topics/post-traumatic-stress-disorder-ptsd/index.shtml.

[63] American Psychological Association, accessed May 30, 2016, http://www.apa.org/topics/ptsd/index.aspx.

which is interesting as the DSM-V no longer characterizes PTSD as an anxiety disorder. The National Center for PTSD struggles to fundamentally define the disorder, stating up front that,

> PTSD can occur after someone goes through a traumatic event like combat, assault, or disaster.[64]

Amidst the extremely large volume of information concerning the disorder, they all speak with relative unanimity on the symptoms that necessitate a diagnosis. The NIMH requires the following,[65]

- at least one re-experiencing symptom (flashbacks, bad dreams)
- at least one avoidance symptom
- at least two arousal and reactivity symptoms (e.g. anger, inability to sleep)
- at least two cognition and mood symptoms (e.g. poor memory, loss of joy)

The National Institute of PTSD lists them a bit differently,[66]

- reliving the event
- avoiding situations that remind you of the event
- negative changes in beliefs or feelings
- hyperarousal

The DSM-V confirms four necessary symptoms: intrusion, avoidance, negative alterations in cognitions and mood, and alterations in arousal and reactivity, seeming to collude with the National Institute of PTSD more so than the NIMH though there is much overlap in the language.[67]

The final criterion revolves around the duration of the symptoms. Most institutions stipulate that everyone will react in some way to a stressor. It is the duration of associated symptoms that differentiates between a diagnosis of an acute response to a stressor or the actual disorder. Most organizations

---

[64]National Center for PTSD, accessed May 30, 2016, http://www.ptsd.va.gov/professional/PTSD-overview/index.asp.

[65]NIMH.

[66]National Center for PTSD.

[67]DSM-V, accessed May 30, 2016, http://www.dsm5.org/Documents/PTSD%20Fact%20Sheet.pdf.

stipulate one month as the minimum duration to be considered for diagnosis though the Mayo Clinic expects months or even years of *worsening* symptoms.[68]

Some and Not Others

A specified pathology is not the intent of this work, only a general understanding of the clinical aspects whereby we may move on to hopefully weightier matters and potentially a more potent comprehension. Why do some respond to stressors in a certain pathological fashion while others do not? What actually drives the response? What drives an experience to transition from stressor to trauma or for that matter, for a singular experience to induce trauma upon one and not the other?

The medical community disagrees on the neuroendocrinology and neuroanatomical conditions associated with PTSD. Studies have displayed that various chemical and hormone levels, such as cortisol and serotonin, are associated with a disordered response to stress. Additionally, brain anatomy and structure, specifically the prefrontal cortex, amygdala, and the hippocampus, all have demonstrated altered function with respect to PTSD. Different types of medications have proven varyingly effective in treating PTSD symptoms, depending upon the source.

An examination of extenuating and mitigating factors provides the most utility for our purposes. The NIMH defines some additional risk factors for PTSD, aside from the actual trauma,

- Physical injury
- Childhood trauma
- Having little or no social support after the event
- Dealing with extra stress after the event, such as loss of a loved one, pain and injury, or loss of a job or home
- Having a history of mental illness

They list the mitigating factors,
- Support network of family and friends

---

[68]Mayo Clinic, accessed May 30, 2016, http://www.mayoclinic.org/ diseases-conditions/post-traumatic-stress-disorder/basics/definition/con-20022540.

- Finding a support group after a traumatic event
- Learning to feel good about one's own actions in the face of danger
- Having a positive coping strategy, or a way of getting through the bad event and learning from it
- Being able to act and respond effectively despite feeling fear

These resilience factors seem to fall into two categories: social and mental. Socially, the existence of a healthy, vibrant, and loving support system of family, friends, and maybe a support group contributes to PTSD avoidance or, in the case of existing PTSD, mitigation of a dysfunctional symptomatic response.

The mental aspects could be grouped under the label of 'self-awareness.' Does the individual have a healthy and holistic view of what actually occurred, of the traumatic event, of his role in the matter? Does the individual have a foundation upon which to rest in times of turbulence? Is the individual self-aware enough to recognize what is actually occurring and to recognize potentially aberrant behavior in themselves before it becomes dysfunctional?

Foundations

In my research, probably the most fruitful encounter was a meeting with Evan Owens, the founder and director of REBOOT Combat Recovery, a non-profit, faith-based program that enables soldiers to understand the roots of trauma, the source of dysfunction, and the path to absolute healing.

This is a spiritual battle, not a battle against the flesh and blood. Many veterans return from battle diminished spiritually and then sink into a litany of behaviors that further exacerbate whatever conditions already exist. I've observed exactly this as issues such as alcoholism, drug abuse, infidelity, and violence rip apart the lives of those afflicted by PTSD. Suicide provides the fullest expression of the conglomeration of dysfunction.

What Owens pointed out to me, what they had discovered in many of their groups, was that dysfunction already existed well prior to the individual experiencing the horrors of combat. Most of those who suffered, as they peeled back the layers, manifested some sort of trauma from earlier times, often from childhood. They had already been traumatized, though it had been unrealized and suppressed. The combat trauma further rendered them vulnerable to a myriad of afflictions and dysfunction.

Most significant of all, most young men today lack a moral compass. They are 2nd generation unchurched. At some point, their fathers either walked away from the church (de-churched) or were themselves unchurched. As such, young men lack the basic spiritual foundation to not only drive appropriate conduct but to comprehend and deal with difficult situations.

> *Their greatest moral injury comes from fathers who either walked out or failed to disciple them, to bring them up in the way of the Lord.*

This generates a spiritual bankruptcy that underscores the dysfunction, a lack of comprehension concerning all things spiritual, a core brokenness. At its center, PTSD springs from the coupling of trauma and this concurrent spiritual bankruptcy rendering the warrior hapless against forces that would seek to destroy him. As with anything, but assuredly concerning the mental aspects of wartime affliction, the metaphysical governs.

Apart from the spiritual, how could we ever do more than treat symptoms and bandage wounds?

Trauma Realized

"Move out," came the order. Conway couldn't believe it. His entire squad had literally been blown up less than an hour before. After evacuating the wounded, SSG Conway was left with less than a team's worth of soldiers to lead the way to the objective. The Platoon Sergeant had evacuated with the casualties and never returned though he was unwounded. "Move out," came the order.

SSG Conway did exactly that, what a soldier does. He gathered his men and equipment and moved to the objective. A short time later, Conway and his men arrived at the building that would provide overwatch of the target building. They cleared it and occupied when they got another call from the company commander. He wanted them to drape a large camo net over the roof of the building to help identify it as a friendly position.

Conway argued that the green camo net against the brown building would essentially make them a target for the enemy, but eventually he relented. Turning to one of his fellow sergeants, they literally did a rock-paper-scissors to see who would climb to the roof and drag the net up. Conway won.

As he sat and watched his fellow NCO climb to the roof he began to feel guilty and he got up to go and help his friend. A small voice in his head warned, "don't go up there," and so he helped from below, pushing the net up to his friend, securing it on the backside. A volley of fire rang out, punctuating the stillness of the night. The sergeant heard a thud and then his friend rolled off the roof and into the camo net Conway was holding. A sniper's bullet had found his neck leaving him grievously wounded but still alive.

He was gently lowered to the floor. "Conway, it's hard to breathe," he managed. Conway called for the medic.

Together they rolled him over and managed to seal the exit wound. Conway knew they had to get him out immediately or he would die. He called for a CASEVAC.

"I can't feel my feet."

A few minutes later, "I can't feel my legs."

Conway slapped him in the face, trying to keep him awake.

"I can't feel anything. It's hard to breathe." Conway again called for the CASEVAC but it never came. It had gotten lost in the city. Finally, after what seemed like an eternity, the vehicle arrived but by this time, Conway's friend was already dead. They had to break down a wall to get the stretcher out. This man had been an impressive and inspirational leader, one who led from the front. He had been Stephen Conway's friend, and now he was gone.

Following the evacuation, all hell broke loose as the entire surrounding neighborhood seemed to unleash upon their surrounded position. For Conway, this was the Alamo. They engaged, and were engaged, for the next several hours from every direction. At some point, they went black on ammunition having fired tens of thousands of rounds. Survival became the objective as thoughts of the original mission evaporated.

Somehow they survived and at last, the gunfire subsided. They pulled out. SSG Conway was now the senior leader in the entire platoon. Following exfil, Conway sat in a daze. His joints, knees and elbows, all of them, began to swell, locking up. The medic gave him some shots for inflammation.

"I don't ever want to come back here," the medic confided in him.

Now in charge, SSG Conway sat and tried to come to grips with this night. "What just happened," he thought over and over again.

"What do I do with this?"

They never did destroy the building, their original target.

Chapter 14

14. Resolution

Standoff

I've watched hours of Kill T.V. over the years, hundreds of hours. It makes for good viewing on occasion. Early in the American Civil War, it was considered fashionable for spectators to observe battles from a nearby viewing area.[69] The brutality and bloodshed of modern war as it evolved soon tempered that practice into non-existence.

Warfare has once more become a spectator sport, this time from 10,000 feet above the battlefield. Modern militaries blanket the battlefield with all manner of drones and observation aircraft wielding high-powered cameras that record the carnage below. They relay their signals to operations centers where commanders and their staffs sit before walls of plasma televisions displaying various feeds from various platforms watching, always watching. One senior leader used to call it the "unblinking eye," a moderately Orwellian term.

Everything is filmed and everything is recorded. Commanders now see everything the enemy does. Various experts monitor these feeds, analyzing and directing assets around the battlefield hopefully culminating in a successful kinetic strike on a particular target, kinetic strike being a euphemism for 'blowing it up'.

As the SDF moved to siege Manbij in the summer of 2016, the allies unleashed a ferocious volley of kinetic strikes, destroying hundreds of ISIS vehicles, fighting positions, and weapons and killing hundreds of enemy

---

[69] John J. Hennessy, "War Watchers at Bull run During America's Civil War," *HistoryNet*, accessed June 3, 2016, http://www.historynet.com/war-watchers-at-bull-run-during-americas-civil-war.htm.

fighters, all directed from an operations center hundreds of miles away. As my task force was working another AO, we found ourselves watching the action from our own operations center as spectators.

"Cleared hot," came the crackled radio call, prompting me to look up from my workstation.

"In," responded the fast-mover.

"30 seconds."

One of our screens displayed in striking clarity an ISIS fighting position manned by a handful of fighters with a mortar tube. They had no idea they were literally less than a minute from destruction.

Boom! A gigantic explosion momentarily darkened the feed as a fireball followed by a cloud of debris and smoke rose into the air.

"Good effects," came the matter-of-fact radio affirmation.

The staff soldiers in my operations center grinned and hooted. "Dang," one of them uttered.

"Only you can prevent forest fires," another grinned.

On the screen, smoke and debris continued to obscure the obliterated target and in effect, obscure the reality of the scene on the ground. This was just an image on a screen, something we'd seen play out countless times. The reality on the ground spoke to a different consideration. Out there somewhere, the mutilated bodies of several young men lay splattered around the blast radius. What had been flesh, living breathing flesh just seconds ago, men with hopes and dreams and lives, was now a vaporized mist.

Now, lest we forget, death is a necessary part of war and a horrid affair, but necessary nonetheless. These men, corrupted by systematized evil, stood as an affront against righteousness. Their death was just. The coupling of technology and tactics effectively eliminated the tragedy of such an affair or rather the obviousness of said tragedy. As we've already stipulated, the destruction of the Image of God is always a tragic affair.

I wondered if the boys would have been so nonchalant had they pulled the trigger themselves, in person. Would they have cheered had they looked these men in the eyes before dispatching them, had they been forced to come to terms with the sheer personhood of the enemy before killing them?

Militaries have always sought stand-off yielding several effects, some intended, some not. The further away I can engage an enemy, the safer it is for my soldiers, pragmatically preserving my combat power. Thus, we seek to engage the enemy from as far of a distance as possible. Modern militaries have nearly maximized this effect. Today, soldiers from bases in the United

States routinely direct drones around battlefields on the other side of the world engaging and destroying the enemy. Strategic bombers deploy from bases on other continents to deliver their payload and then return to home base. A soldier literally might leave home for work, fly to a distant land and slaughter several hundred enemies, and be home in time for a late dinner. Not only does this serve to protect our own soldiers, it provides the benefit of increasing the likelihood that a soldier will engage.

Operant and classical conditioning have proven effective in overcoming a man's natural inclination not to kill. Standoff yields a similar result. Thus, a soldier will more readily shoot an enemy from 100 meters with a rifle than stab him in the gut with a bayonet. A soldier will more readily obliterate an enemy fighting position with a mortar from 1000 meters than shoot him with a rifle. A pilot will more likely engage from 10,000 feet than the mortar man. I've never heard of a soldier refusing to kill via remote link, through Kill T.V.

Many thought that increased standoff would yield the added benefit of mitigating trauma, the trauma induced from the destruction of the Image of God. Operant and classical conditioning increase the effectiveness of the individual soldier in killing, but render the soldier vulnerable to trauma along with all of the potential side effects that often accompany trauma. However, the proliferation of remotely executed warfare has given rise to a recent phenomenon, the increasing proliferation of mental health issues among those who practice remote warfare exclusively, a seemingly counter-intuitive proposition. Drone operators are being diagnosed with PTSD in increasing numbers.[70]

On top of this, several recent studies concerning the military and suicide generated some surprising results. A majority of those who attempt suicide have never even deployed to combat, many being new soldiers in their first few months of service which speaks to a general spiritual bankruptcy I'll address in Chapter 27.[71] Combat generates trauma and that sometimes tragically manifests itself as suicide. Yet, trauma proliferates within many disparate populations. Perhaps the other factors associated with trauma

---

[70]James Dao, "Drone Pilots Are Found to Get Stress Disorders Much as Those in Combat Do," *The New York Times*, accessed June 11, 2016, http://www.nytimes.com/2013/02/23/us/drone-pilots-found-to-get-stress-disorders-much-as-those-in-combat-do.html?_r=0.

[71]Sarah Childress, "Why Soldiers Keep Losing to Suicide," *Frontline*, accessed June 10, 2016, http://www.pbs.org/wgbh/frontline/article/why-soldiers-keep-losing-to-suicide/.

prove causative.

Despite our best attempts to eliminate the occurrence of the actual event that might generate trauma, incidents of trauma and associated dysfunctional responses continue to proliferate, becoming even more widespread. As we've seen, a singular event may prove traumatic for some and routine for others. One soldier may be able to shoot enemies in the face with impunity and never lose a minute of sleep while another may suffer from dropping bombs remotely or from the sheer presence of danger, the potential for trauma proving traumatic in and of itself. The event is what it is. It is the response we must shape.

Though we deny the secular, an examination of the mitigating factors associated with secular PTSD research proves fruitful, though we will need to dig a bit deeper. Three factors serve our purposes in shaping responses to events that might otherwise prove traumatic:

1. Socialization.
2. Clarity.
3. Forgiveness (turn to Section 4 for a discussion on this).

Band of Brothers

Henry V stands before his soldiers shortly before the Battle of Agincourt and declares,

> We few, we happy few, we band of brothers;
> For he to-day that sheds his blood with me
> shall be my brother.[72]

The profession of arms is a special profession. The brotherhood of the profession binds men unto one another. It is the brotherhood that unites all soldiers, a bond forged in the unique fires of the crucible of bloodshed and corporate conflict.

Aside from group norms and group dynamics, men appropriately bound find security in solidarity. As I march lockstep with the man on my right and left into the fray, I march with the solidarity of a shared purpose in confrontation of a shared enemy. No matter what demon might seek to

---

[72] William Shakespeare, *Henry V*, accessed June 13, 2016, http://quotationsbook.com/quote/3024/.

traumatize me, I find solace and comfort in the sharing of this burden with men of the same cloth and caliber. At some point, the brotherhood supplants the mission; it becomes the mission.

I can think of no other institution that evokes a fellowship, a brotherhood the way that martial endeavors do, other than the Church itself. This brotherhood, aside from emboldening men to accomplish that which they might not otherwise be able, serves to steel men's nerves against affliction. If I share the burden with my brothers, then the burden becomes that much lighter.

One of the main factors that generated intense mental trauma from Vietnam was the manner in which they deployed. The units remained deployed and were largely filled with an ever-rotating pool of temporary hires. Units fought with hired guns essentially. There was very little unit integrity and fidelity. It would be very similar to a football team playing each game with new recruits. The team would never develop any team chemistry or unity and would likely be ineffective. This had the same effect in Vietnam and diminished the soothing effects of the brotherhood. Brotherhood existed in spite of the system, not because of it.

Is it any wonder that soldiers miss war? How could we ever replicate the brotherhood in civilian life? SSG Conway confided to me that this was the main aspect of military service he missed, the comradery. Before, he had his band of brothers, his comrades, those who had shared the same experiences, eaten the same dirt, shed the same blood, and bled from the same wounds. Once the Army turned him loose, he had nothing, or so he perceived.

The need persists, perhaps to an even greater extent once a warrior departs the company of his brothers. In this case, community becomes essential, family vital, and the Church critical.

How is the warrior supported? How are the functions of the brotherhood replicated once a warrior departs the ranks? The potential always exists for an event to become trauma and to translate into dysfunction even years later. The community must necessarily assure and reassure. The family must remain strong and we must surround the warrior. We must, in essence, rebind the warrior in the love of the group, multiple groups, knowing that his participation in the most difficult of tasks leaves him vulnerable to intense spiritual affliction.

A warrior unbound might just become a warrior unhinged.

Clarity, part one

"Issue the order."

Master Sergeant Roddie Edmonds stood defiantly before the Nazi commander. The Nazi commander pressed the barrel of his pistol into Edmonds' forehead, "Issue the order."

Edmonds was captured alongside thousands of others in late 1944 at the Battle of the Bulge. The Nazis imprisoned him in the Ziegenhein Stalag where he spent the next 100 days in captivity before being liberated by the advancing Allies. It was at Ziegenhein that Edmonds was put to the test, confronted with the unspeakable horror of the Nazi Final Solution to the Jewish problem.

Allied commanders had long advised Jewish soldiers, in the event of capture, to hide or remove any evidence of their Jewish heritage. By the winter of 1944, Nazi Germany closed most of the notorious death camp; the Allies liberated others. They still sent Jewish prisoners of war to labor camps where the combination of forced labor, minimal sustenance, and horrific living conditions caused the death of untold thousands of POW's. Unto this, the Nazi commandant at Ziegenhein demanded that Edmonds, the senior ranking soldier, order all Jewish soldiers to present themselves the following morning for formation, intent on moving them to a labor camp.

At that instant, Edmonds was confronted with a choice, a choice with moral ramifications concerning the collective good versus individual benefit, a choice with truly life and death implications. Having fought at Bastogne, he certainly experienced potentially traumatic events. Capture itself produces a unique form of trauma known as *capture shock*. During several months of German captivity, Edmonds likely witnessed and directly experienced potentially traumatic events. Now, he stood tested, presented with a wicked problem. His decision would likely resonate in his mind and soul for years to come.

Rightness of Mind

To assess what we must do with all of this, we turn to the one place guaranteed to provide answers.

Trust in the Lord with all your heart, and do not lean on your own understanding.
Proverbs 3:5

Most people rely upon their own intuition or the counsel of others. This issue is that our intuition is skewed by a number of factors, mainly our sin, and much in the way of external counsel runs absolutely contrary to the word of God.

I had one of my sergeants come to me once with marriage trouble. He and his wife could not get along and were contemplating a divorce. Having attended some marriage counseling where the counselor had offered some unhelpful advice, he confided in me, "I wish someone would just write down the answers in a book or something." I almost fell out of my chair. This guy professed Christ. I gestured toward the Bible on my desk and raised my eyebrows, looking at him in disbelief. "Yeah, I know, but not that…you know what I mean." During difficult times, we must take extra care in where we find counsel.

Many issues originate in a perceived inability to govern our thought life. Scripture speaks to the opposite.

> *"For who has understood the mind of the Lord so as to instruct him?" But we have the mind of Christ.*
> *1 Corinthians 2:16*

The believer is equipped with the literal mind of Jesus, endowed by the Holy Spirit. Paul further exhorts us to, "take every thought captive to obey Christ." (2 Corinthians 10:5) I possess the ability to control my mind, to govern my thoughts, to take them captive to the obedience of Christ. I may discipline my mind and enslave it to Christ.

It is a difficult thing to empty one's minds of something, to eliminate that which is harmful from one's thoughts. It can be done, but effectively, I'm not sure for how long. Scripture calls us to the contrary, to *fill* our minds with that which is helpful and in this way, we may truly govern our thoughts. Paul again exhorts believers,

> *Finally, brothers, whatever is true, whatever is honorable, whatever is just, whatever is pure, whatever is lovely, whatever is commendable, if there is any excellence, if there is anything worthy of praise, think about these things.*
> *Philippians 4:8*

We should dwell upon these things. We ought to fill our minds with godly thoughts, with the word of God, and all things true, honorable, just, pure, lovely, and commendable. Paul exhorts us to think about these things. One verse prior, he reminds the Philippian believers about the "peace of God, which surpasses all understanding" and that it will "guard (their) hearts and minds in Christ Jesus." (Philippians 4:7)

Millennia earlier, Isaiah encourages,

*You keep him in perfect peace whose mind is stayed on you,
because he trusts in you.
Isaiah 26:3*

Clearly, peace of mind closely relates with fixing one's mind upon the Lord Jesus. What of the event? As we discipline our minds, fixing them upon all things godly, at some point we must reconcile the event with our position before God.

We've examined certain aspects of an event that might induce a traumatic response but there exists endless possible variances. We must recognize that all things we experience in life help to shape our character, good or bad, positive or negative. Our very being exists as a summation of confluent factors including life experiences. War changes men, inevitably, assuredly. Through the conduct of warfare, men experience horrific events including the taking of human life or maybe even the witnessing of, or participation in, atrocity. War often causes significant changes in a man. How does one keep from being defined by the event, from allowing the event to manifest itself as dysfunction replete with the label of trauma? God never intended men to live governed by trauma.

What Do I Do With This?

Time is an issue. Conway's question is so poignant in its simplicity and absolute *necessity*. "What do I do with this?"

*I've killed in war, what do I do with this?
I've witnessed or participated in atrocity, what do I do with this?
I've seen friends die, now what do I do with this?
I don't understand why, but what do I do with this?*

Time becomes the enemy of reconciliation. In the modern army, soldiers are frequently shipped straight from combat to their living rooms. They have little time to consider, to ponder, to discuss, to think, to reconcile.

On the night that our task force killed AMZ, we were scheduled to RIP out. As it was, we executed several operations that night, met our replacements on the tarmac as we landed at sunrise, and then promptly boarded the Air Force aircraft for the flight home complete with a crew swap in Germany further expediting travel. As me and the Gunny boarded the aircraft we remembered that we hadn't eaten in many hours and there was no food and we were starving. I distinctly remember digging through the garbage from the inbound flight to find some bits and scraps.

It was more than a little unusual, relaxing in the cargo hold of the C-17, watching the loadmaster rest his foot on the cooler that unbeknownst to him contained a DNA sample (body part) of AMZ that we were transporting to the states so that the authorities could positively identify the body. I popped some Ambien and a few minutes later began throwing up all over myself, rushing to the bathroom before any could notice. Less than 24 hours later I was walking through my front door, back to the real world.

SSG Conway's return from his second tour in Ramadi was very similar. In many instances, soldiers were pulling guard duty and fighting up to the day that they got on the plane to fly home to reunite with friends and family. They have no time to consider these deepest of questions concerning God and the thing that is war. They have no time to process and answer the question, "What do I do with all this?" They are expected to reconcile on the go and walk back into the real world without a glitch. Is it any wonder that many struggle?

In the conclusion of Grossman's spiel, he likens a traumatic response to a puppy let out of its kennel. It gets out, we acknowledge that the puppy is there, and then we put it back in its kennel and move on with life. Part of my intent is to provide you, the Warrior, with a proper view concerning the thing that is war and its relationship with the individual actions of individual soldiers.

We pursue moral clarity. If I have the moral clarity to recognize and acknowledge the existence of an evil that necessitates confrontation and that the Lord calls men to be strong and courageous in confronting this evil, then I can reconcile my own actions with this knowledge and better recognize where I stand before a holy and righteous God. If I can erase the moral vacuity that pervades in the minds and hearts of many young warriors, I can more readily recognize the puppy when it emerges from the kennel and ever more dispassionately put it back into the kennel. Eventually, as I discipline my mind concerning proper things, the puppy may not reappear as often or even at all.

More on Scarring

The puppy is still there and *I want it to be*! The existence of the puppy implies respect for the dignity of human life. Were I to participate in the taking of human life with absolutely no ill effects whatsoever, that would be cause for concern. We must acknowledge the existence of the tragedy that is the destruction of human life and the associated incorporation of this into war. We can never lose this. Let us never be comfortable with the notion of killing.

As the conduct of war demands strength and courage, the termination of hostilities places similar demands. All killing scars the soul a bit, even righteous killing for a righteous cause enslaved to a righteous God. We know this yet God calls us to fight just the same. We participate with the full knowledge that the act of killing will affect us. God called us to have strength and courage, during war and maybe more importantly, after.

Let's make it personal. Our current enemy embodies evil and God has called men to stand in confrontation. I know that the act of killing will affect me in some way yet because as a man, I subordinate myself for the needs of my wife and children, I am willing to bear that burden on behalf of them. The strength comes in willingly bearing the scars of war on behalf of those I love and defend. With each death, I give away a bit of myself which is okay. That's a sacrifice God calls the warrior to make.

This must be part of the reconciliation, call it an incomplete reconciliation if you will. I understand and reconcile that which God has called me to and embrace the given peace. I reject dysfunction as an acceptable and necessary by-product though I never completely erase the effects from the destruction of human life, nor do I want to. That would, in essence, deny what I so desperately seek to preserve: my humanity.

Binding the Church

To the Church we turn for the answer. Too often, especially in contemporary western society, the Church willingly relegates itself to the backseat while the government or other organizations deal with a particular issue. It is with much shame that our society requires a vast government bureaucracy to handle the glut of orphans in the country. The Church should lead the way in caring for the orphan, in providing loving, godly homes for children who have no parents or homes.

Much the same could be said of the Warrior and the one who laid down his sword, yet still lives afflicted by the horrors of war. The Church ought to lead the effort in binding them, by enveloping them in the loving bonds of community, family, and fellowship. The Church, as the body of Christ, is the only institution that can replicate the brotherhood of arms, even surpass it in depth of intimacy. I have become, over the last several years of immersion in my church, intensely tight with a new band of brothers, most of whom have never served in the military.

Support groups are good. Non-profit institutions serve a purpose as well and the government even has its place but the Church, the Church ought to be in the business of seeking the lost and bringing to them the words of life and Christian community. The leadership of REBOOT Combat Recovery understands this paradigm as they have sought to lash their organization with the local church. They desire that their organization serves as a gateway to immersion in the Christian community for the struggling warrior, a truly noble endeavor. They offer the warrior a program, but really offer Christ through the Gospel and subsequent immersion in a local body of believers— the only path to true healing.

Clarity Part Two

"Issue the order," the Nazi commandant demanded.

Edmonds looked to his left and to his right, hesitated, and said to his fellow POW's, "We are not doing that, we are all falling out." Every single man presented himself before the commandant. As all of the inmates stood in defiance, the commandant said to Edmonds, "They cannot all be Jews."

"We are all Jews here."

I can imagine the disbelief in the commandant's heart, that this man would brazenly defy him. I can feel the solidarity in the hearts of the men as they boldly stood against the wickedest of evil, possessing the moral clarity to understand that they faced a truly corporate evil and that the *only* response was a unified band of brothers refusing to yield. The commandant placed his gun to Edmonds' head, repeating the demand. Edmonds offered his name, rank, and service number as required by the Geneva Convention.

"If you are going to shoot, you are going to have to shoot all of us because we know who you are and you'll be tried for war crimes when we win this war."

The audacity of Edmonds, the boldness and courage, imputed directly to his brothers' hearts. They fed off this single act of rebellion against evil and united as brothers.

After the war Edmonds never spoke of the encounter. Decades later, his son who had been perusing some of his dad's war journals began researching the incident and discovered the truth. A Baptist pastor, Chris

Edmonds knew it was the deep moral convictions of his father's Christian faith that allowed him to triumph at his hour of testing. That it would be the same with you, Brave Warrior.

Reconciliation

I don't know that Stephen Conway ever fully reconciled. His boss at recruiting command sent him to a one month inpatient treatment at Walter Reed where, for the first time, he began to come to terms with the fact the there was a problem or an issue of some kind. They did the typical: pumped him full of numerous medications including a battery of psychotropic drugs, offered him counseling and even some near-eastern types of homeopathic treatments. It seemed to help, enough that he was able to deploy with the 101st.

The first part of that tour went well enough until the single round to his forearm brought back all of the memories and fears and anxiety. That single AK round unleashed the puppy from the kennel in a definitive way.

He lived, medically retired from the Army, and eventually found his calling as a high school football coach. He pours himself into his family—he now has a daughter to go with his son—and his boys on the team, investing in them, coaching them, teaching them. He sees in them the unmarred and untapped potential of youth.

God, in His providence, began to answer some of Stephen's questions. When he was a platoon sergeant, one of his soldiers was actually an ordained pastor. Again Conway asked, "What does God think about me killing people?" This time, the man of God had an answer. He pulled out the Bible and explained to Stephen what it said.

"Do you have any regret," I asked.

"No. I brought my soldiers home. I did what I had to do and brought them home."

After a pause, he continued, "You are tainted though, and nothing will ever take that away."

At some point, Conway returned to the church. "My faith is stronger than it was before." He found comfort and compassion in the military community surrounding his army base, through the military members in his church. He found a rekindled excitement for the faith, and he also got off of every medication.

"What would you like to tell any soldiers reading this," I asked.

Conway thought for a minute. "Secular treatment is patchwork. It works temporarily but it never gets rid of the problem. I was in my darkest moments when I was spiritually inactive, a loan wanderer. I decided to get plugged back in out of necessity and my symptoms diminished. They will never go away completely."

A full reconciliation, though always sought, is an unattainable dream. The real sacrifice of the soldier is the acknowledgement of this burden and a willingness to bear it on behalf of those we love. Pray that you might bear it well.

Section 3:
War and the Heart

Chapter 15

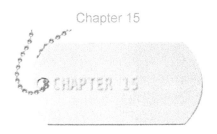

Wednesday, May 4th (Layover)

I've got about two hours until we have to get ready to go again. The system fully engulfs us. We lack any control. Some will rage against the system on occasion, always fruitlessly, leaving the subject spent and even more frustrated. Almost always, whatever condition yielded the meltdown still persists.

I recall a particular incident whereby the security folks at the military airport seized one of our pilot's knives. He had inadvertently brought it in his carry-on to the terminal and as we proceeded through security a short time later, they confiscated it. This was the wrong man to mess with. We call him "Tourette's" for a specific reason. A soft-spoken, humble man, and a great pilot, every now and again he would quite literally blow his stack and go off. It was not a pretty site.

As I used to crew with him quite a bit, he is also the only man I know who has ever successfully taken a crap in the front seat of a Blackhawk helicopter...while fully kitted up...in flight...at night...just prior to an aerial refueling plug. I am still unsure as to how exactly he pulled off this miraculous feat of airmanship, but I digress. They took his knife and he wasn't happy.

His points were all valid. We headed home from combat having just completed yet another busy tour in Iraq. Our C-17 was loaded to the gills with weapons and ammunition and all manner of the tools of the trade and they were worried about...his pocket knife. It was a nice knife, brand new.

He erupted.

He raged.

He informed them that this was a load of bull****.

He demanded to see the General.

He informed them that this policy was a bunch of liberal garbage.

He ranted.

He raved.

He nearly blew a gasket and then he conceded defeat and boarded the plane…without his knife. They took my brand new tube of Crest 3D white toothpaste as well. It is hard to describe the frustrations of the system, the invisible web of policies and procedures that exists to move large volumes of men and equipment around the globe as efficiently as possible with little regard for comfort and/or convenience.

Just now, I went to get a bite at the chow hall but realized I left my ID card at our temporary barracks. I had to surrender my ID card to borrow a towel and had not yet turned the towel back in to retrieve my card necessitating a 20-minute delay in chow as I walked back to the barracks, secured my towel, turned it in for my ID card and then trudged back to the chow hall.

You have to submit to the system. We will emerge in Iraq within the next 24 hours or so. It's as inevitable as the rising of the sun.

I'm pondering life and death and those who would go. What is it that makes men confront death in a certain manner? ISIS shot up the boys on the last few missions. By the sheer grace of God, they failed to bring a helo down, though a couple returned from the mission with several bullet holes necessitating some lengthy repairs. They hit a couple of the ground force as well, nothing too serious but the brass had started to notice.

As much as audacity enables victory, complacency in previously successful audacious endeavors will yield disaster at some point. Each foray against the enemy must be considered on its own merits. Just because something worked previously it may not work again, even if all conditions are still the same.

The group we were set to replace had re-learned this hard lesson. My number one priority for this assignment is to bring these men home to their families.

"What about victory?" you might ask. Well, that is priority 1B and we'll do what we always do and get after the enemy. Yet, for every bad guy we kill, three more take his place. It takes years to create and train one of these men, and for their families there is little solace in the honor of sacrifice.

I pray that we will fight fiercely, with honor and pride, and that the Lord will hand our enemies over to us in victory. I pray that the Lord will see fit to allow every one of these men to return safely home, honoring Him as we go. May the Lord be with us.

15. Hearts on Display

A Switch Flipped...

Overt dissimilarities may not exist, in speaking of the conduct of war with respect to those who conduct it. As I served and continue to serve, as I prayed and dwelt, I inevitably moved beyond questions of 'ought' and 'should' as in, 'ought a Christian to fight?' I don't remember exactly when it happened. There was no decisive point, no precise moment of clarity and illumination. It just happened and I found myself dwelling upon questions of 'how'. How should a believer conduct himself during the course of war?

On the surface, the believer might wage war in exactly the same manner as the heathen. Do internal differences matter?

I recall a particular mission. We were preparing to rush out the door to execute a hasty objective that had presented itself somewhere in central Iraq. As we marshaled, conducted a hasty briefing, and headed for the door one of the crew chiefs looked around and announced with a casual smile, "Let's go kill some Muslims," projecting a definitive eagerness at the task before us.

Now, in all likelihood, that is precisely what was going to happen. The intended target was likely a Muslim and at the time, many were choosing death over captivity though sometimes we gave them an option. Sometimes we didn't. I don't remember on this particular objective which was the case, but there remained a very high chance that by our hand, the actions of our combat team, a Muslim would die.

It is also a fact that this particular soldier's actions and mine would be indistinguishable from one another except for the specific aspects of our different occupational specialties. Combat is a team sport and we were members of the same team.

In reality though, neither of us would likely pull the trigger. As a crew chief, he would conduct his crew duties in assisting the pilot in maneuvering the helicopter toward the intended target—an intensely vital duty. As a pilot, I would either maneuver the helicopter or assist my co-pilot in maneuvering the helicopter. A sniper from the rear or one of the ground force would likely fire any necessary shots. However, crew chiefs and door gunners did kill Muslims—many of them became excellent shots with the door gun—and gunship pilots have also slain many. It's just that in all likelihood, this particular soldier would probably not kill any Muslims on this particular day.

It was pure bravado. I recognized that. This soldier was a member of a team that would possibly kill a Muslim in the coming hours and he expressed his matter-of-fact confidence on the issue. Again though, apart from our individual crew duties, his actions and mine would be indistinguishable. We were both part of a team that would possibly kill a Muslim in the coming hours.

I could not affirm his comment with a hearty, "Yes indeed, let's go kill a Muslim!" I pretended not to hear, brushed it off, and went about my business though the thought stuck with me for some time.

Were his actions different than mine, whether either of us actually pulled the trigger or not? Did it matter that I felt differently? I too possessed an eagerness, a desire to close with and destroy the enemy. However, my thoughts were not quite congruent with his.

I've heard soldiers speak about their switch. Earlier I talked about the switch I discovered at some point, when bidding farewell to my family, switching from familial love and sadness at departure to the 'business' of warfare. I'm not referring to that switch. I'm referring to the dichotomy that some soldiers seemingly maintain in their lives with respect to the conduct of warfare versus their conduct elsewhere, perhaps at home or in everyday life.

I had a soldier confide in me to the existence of his switch whereby he goes from 'family' mode to 'combatant' mode as these two existences are not compatible or consistent with one another at least in his particular case. Ostensibly this soldier and likely others *act* differently during the conduct of war than they do at home. They must, in essence, become something else, somebody different, in order to effectively conduct their wartime duties.

I ask, is this necessary? Is this biblical? I wondered if I should have a switch such as this.

The Warrior Ethos

In wondering how I, a believer in Jesus Christ, should act during the conduct of war, I considered that the Army has already told us. In November 2003, Army Chief of Staff General Peter Schoomaker approved the *Warrior Ethos* as a part of the *Soldier's Creed*. It states,

> I will always place the mission first.
> I will never accept defeat.
> I will never quit.
> I will never leave a fallen comrade.

These are just words on paper. Every soldier says these words. How many actually live them?

As we've examined the corporate aspect of war, we'll turn inward toward the individual though we'll not abandon the organizational. We'll return to it in much detail later, but for now we will drill down through collective motivations, down through national level objectives, organizational aspects and seek to answer questions as they pertain to the individual soldier.

The individual soldier stands ready, brandishing his weapon, bound by the invisible organizational restraints of his unit and other invisible restraints such as the Laws of Land Warfare and Rules of Engagement, confronted by the reality of a flesh and blood enemy, faced with a clear choice. He must decide how to act, how to conduct himself, what he should do, and he must reconcile his actions or potential actions with the firmest convictions of his inner being.

Thus we'll seek to understand and answer these questions. In what manner should a soldier conduct himself? In what fashion should a Christ-follower conduct himself during the course of war? What of the leaders? What about those responsible for the motivation of others? Are these even the right questions?

An Ethos Lived

May 14, 2006 dawned with bright heat, just like any other day in Iraq. The sunbaked desert sands stretched to infinity, the thermometer climbing well above 100 degrees. Mother's Day.

The operation started as a vehicle takedown. While my task force slept—we were on a night cycle in another part of Iraq—another task force hunted and found themselves embroiled in a battle for their lives.

The task force patrolled the western Iraqi desert expanse for years, seeking a fight, seeking the brass ring, a militant brave enough or ignorant enough to venture into the open desert. From Baghdad to Al-Qaim, the blood-stained sand testified to the ferocity of conflict as the former stomping grounds of Abraham brimmed with strife.

I've tried to imagine being in the target vehicle as the assault went down. At what point did the occupants realize what was happening and say to themselves, "This is actually happening?" Though we drew first blood, the enemy never wavered no matter the overwhelming force applied or our team's audacity and agility. A vast number of them proved ready to die and our task force was more than ready to assist them in that regard. We just didn't want them taking any of our guys with them.

As invincible as we felt, we operated well aware of our vulnerability. A vehicle takedown gone awry could quickly deteriorate into an untenable situation. Such was the case on this particular day.

Mother's Day 2006 proved a great and terrible day. It was a tragic day but a day when men rose to the occasion, when hearts full of love and fellowship prevailed. On May 14, 2006, we witnessed the triumph of the warrior spirit, the fullest display of the warrior ethos as pledged by all soldiers—courage and love manifest in the hearts of men. A singular objective drove the spiral, kill AMZ.

The Triangle of Death

The task force hunted the Jordanian Abu Musab al-Zarqawi (AMZ) for several years. He proved extremely difficult to corner. Man-hunting is tough business; finishing proved easy in comparison. It was the finding and fixing that challenged and even drove the reshaping of the entire task force.

Following the invasion in 2003, AMZ rose as the leader of Al Qaeda in Iraq (AQIZ), the champion of the insurgency. His methodical brutality stunned everyone as he targeted Shia Muslims, apostate Sunnis, Iraqi government, and American military forces with equal aplomb. He openly sought to foment a sectarian civil war between the Sunni and the Shia and were it not for the Sunni Awakening in 2007, he may have succeeded. The savagery of his tactics was not without its detractors even among his supporters as even the Al Qaeda leadership, at one point, reprimanded him for the brutality of his methods.

AMZ staged the first videotaped beheading in 2004, the Nick Berg video. The masked man who read the statement and slaughtered Berg was allegedly AMZ. He was later implicated in another American beheading along with a series of bombings, assassinations, and other atrocities. By 2005, he had established himself as the most wanted man in Iraq which is where our task force enters.

The Sunni Triangle, the area between Baghdad, Balad, and Ramadi in central Iraq became a hotbed of insurgent activity. Yusufiyah in particular buzzed with resistance as in many areas, the insurgents operated with complete impunity. Operations in these areas always entailed risk and almost always resulted in a gun fight. This day was to be no exception.

Before the target vehicle could be taken down, the occupants fled and strong-pointed a nearby small compound. They barricaded themselves in, and prepared to defend in place. The ground force had them pinned down when the entire neighborhood erupted. Gunfire poured in from every direction, insurgents maneuvered on the flanks, the sky came alive with AAA fire, the radios blazed to life.

"Troops in contact! Troops in contact!"

The assaulters, all experts at ground warfare, sought cover and directed several Air Force fast mover engagements, dropping one 500lb bomb after another. The operation quickly deteriorated from the original takedown to an all-out gun fight. The insurgents must have sensed vulnerability as hundreds of them closed on the area. The attack helos overhead swarmed the enemy, ferociously and fearlessly laying down suppressive fire to protect the ground force.

Effective enemy ground fire brought down one gunship and then another. Miraculously, the pilots deftly maneuvered them safely to ground right in the middle of the gunfight and right in the middle of the gunfight, the pilots replaced damaged rotor blades, reloaded ammunition, and launched back into the fray.

Forty miles away at the headquarters, my friend who was in command issued the order to alert the night crews. Things were getting serious. I can only imagine the tension in the operations center, the fog of battle, the struggle to make sense of the developing situation.

Valor Anticipated

Though I wasn't there, I easily envision what happened next as Major Matt Worrell, Chief Warrant Officer 5 Jamie Weeks, and their lead crew were woken from a deep sleep. Downrange, sleep is the top priority and pilots normally lived in little metal boxes called CHU's (Containerized Housing Units). In contrast to the bright desert day, the insides were pitch black and ice cold.

A sharp knock at the door as the runner relays to them that troops are in contact (TIC) and they are needed at immediate level 1. The rush of adrenaline as they quickly don their uniforms that they keep at the ready, secure their weapons, grab a protein shake and a bottle of water on the way out the door, and sprint to the flight line after a hasty brief. Within a few minutes, they are kitted up, strapped in, bringing the aircraft to life.

With the flip of a switch, AC power surges through the circuitry bringing the aircraft on line. As the radios awaken, the confusion of battle greets their ears. Quickly, they bring their team to level 1 and notify the operations center, await instructions, listening to the voices of their brothers-in-arms fighting for their very lives. I envision very little conversation in the cockpit as they pull pitch and point their aircraft west toward the sound of the guns.

16. The Theoretical Heart

On the Motivations of Men...

Since men first arranged themselves into fighting organizations, into armies, we've sought to determine what motivates men and collections of men. Theorists, generals, military leaders, and historians have all sought to diagnose motivation. We exalt historical instances of battlefield heroics and scrutinize accounts of wartime failure, oft attributed to a failure to motivate.

Military scholars and practitioners often attribute motivation to leadership, as a function of effective leadership. General Dwight D. Eisenhower defined leadership as,

> The art of getting someone else to do something you want done because he wants to do it.[73]

Eisenhower defined leadership as a matter of motivation, that is, inspiring soldiers to the point whereby your motives become their motives. The leader imputes his motives to his soldiers whereby they adopt or internalize what the leader finds important and make it their own. They have ownership.

This idea possesses much merit and utility. However, many leaders today lack precisely this: the fundamental ability to inspire, usually because they themselves are not inspired by anything higher than themselves. Men want to be inspired. Soldiers want to be inspired. They respond to the higher call, most of them.

Though valid, attributing motivation to leadership fails to address

---

[73] "Leadership According to Eisenhower," accessed May 13, 2016, http://leadership.w9z.org/leadership-according-to-eisenhower/.

fundamental sources. What about the leader? From where does the leader find his inspiration, his motivation, were he to seek to impute it to his soldiers?

We seek insight from debacles such as Task Force Smith, the Chosin Reservoir, or Dien Bien Phu. We assess and analyze the orders and actions of leaders like LTC Charles Bradford Smith (no relation), LTC Don Faith, and Col Christian de Castries. We exalt manifestations of courage, bravery, and selflessness and revel in heroic tales from battles fought, won, and lost.

At the same time, we look to the actions of the enemy, still assessing motivation. We seek to understand how the pilots of the *Divine Wind*, the nearly 3,000 Japanese *kamikaze* pilots, so fearlessly plowed their aircraft into Allied ships desperately trying to turn the tide of the war and avert an invasion. We marvel at the crass disregard for human life of the Red Army at Stalingrad and the ruthless efficiency of the Wehrmacht in overrunning Poland and France. What would possess ISIS militants to drive explosive-laden bulldozers into Peshmerga defensive lines?

Eventually we necessarily turn to a darker subject. What of atrocity? In Vietnam, Lieutenant Calley, Captain Medina, and the soldiers of C Company, 1st Battalion, 20th Infantry Regiment slaughtered at least 500 unarmed civilians, including women and children, at My Lai. At Dasht-i-Leili in December 2001, Northern Alliance soldiers under the Afghani warlord Dostum jammed nearly 7000 Taliban prisoners into metal shipping containers. Most suffocated to death, survivors were shot. In July 1995, Bosnian Serbs raped and massacred more than 8,000 civilians. While U.N. Dutch peacekeepers looked on, young men were rounded up and shot, babies were beheaded, little girls were raped, and people were buried alive.[74] How may we account for these things?

How do we reconcile the soldier who gives his life to save his fellow man with the soldier who commits unspeakable acts of violence and terror against that same fellow man? What motivates men to behave in either extreme, heroic in selflessness or horrific in atrocity?

What of the Heart...

Biblically, it is the heart that drives us and all that we do. Referenced over 1,000 times, the heart is the most common anthropological term in the

---

[74]Graham Jones, "Srebrenica: A Triumph of Evil," *CNN*, accessed May 13, 2016, http://www.cnn.com/2006/WORLD/europe/02/22/warcrimes.srebrenica/.

Bible.[75] Never referring to the physical organ that pumps blood, the heart is the seat of the will and emotion, the source of cognition, that which governs us. We might consider the heart the embodiment of the immaterial or the link between the metaphysical and the physical as what we conceive in the heart inevitably comes to pass.

Strong's defines the heart as,

> the seat and center of all physical and spiritual life,

and,

> the soul or mind, as it is the fountain and seat of the thoughts, passions, desires, appetites, affections, purposes, endeavors.[76]

As the seat of the will, Paul and Luke both assign decision-making capacity to the heart.

*Each one must give **as he has decided in heart**...*
2 Corinthians 9:7a

*Why is it that you have **contrived this deed in your heart**?*
Acts 5:4b

*But thanks be to God, that you who were once slaves of sin have **become obedient from the heart**...*
Romans 6:17

As the source of the will, all decisions originate from the heart. The Old Testament actually combines the functions of the heart with what the New Testament attributes to the mind. Thus, in some respects, cognition resonate from the same place as our passions, motivations, and will.

All moral conditions spring from the heart. As the seat of the will and emotion and decision-making, the heart governs who we are. Our actions, our thoughts, and words provide an indicator of the actual condition of our heart.

Conceptually, a proper understanding of the heart finalizes the destruction of *free will*. In chapter 3, we discussed free will in depth and

---

[75]"Heart", *Baker's Evangelical Dictionary of Biblical Theology*, accessed April 18, 2016, http://www.biblestudytools.com/dictionary/heart/.

[76]"Heart," *Strong's Concordance*, Biblehub.com, accessed May 13, 2016, http://biblehub.com/greek/2588.htm.

determined that there is just no such thing as ungoverned decision-making capacity. Free will exists, but the will is *always* enslaved to something. It is enslaved to our nature as a function of the condition of our hearts. The heart rules over the will.

Wicked by Default

Some argue for the inherent goodness found in men. They postulate the basic goodness of all men and even more so, their own general goodness. The latter conclusion is most desired, the former merely being necessary to arrive at the latter. When someone says that all people are basically good, what they normally mean is that they themselves are basically good. I had two interesting encounters at a city festival the other night which highlighted this misbelief.

My friend had set up an evangelism booth and we were sharing the Gospel with whoever might listen. At one point, two young men and a young lady rested on the curb opposite our booth for some time as the huddled masses ambled by. After a bit, I went over to speak with them and asked if they minded if I shared the Gospel. They agreed, a bit hesitantly, and so I squatted down and began. Once we entered the meat of the Gospel it went a bit like this…

Me: "Are you a good person?"

Them: "Sure."

Me: "By whose standard?"

Them: "Subjective morality?"

Me: "You mean your own standard?"

Them: "Yes."

These young folks were perfectly at ease with measuring themselves by their own subjective standard. After all, if you can set the standard, define the terms, then you will never find yourself wanting.

A great many in the west believe in this way, acknowledging their own inherent goodness based upon a fluid, subjective, and ill-defined standard. No one ever told them differently and to consider themselves up to standard, no matter which standard they apply, is a perfectly comfortable proposition.

A century ago, a different dilemma, a different paradigm confronted preachers—that of convincing men that God could love them. So convinced were people of their sin, of their wickedness, that preachers' primary task was convincing the people that yes, in spite of your wickedness, in spite of your sin, God still loves you.

Today, the tables have turned. The script has been flipped. Today, if you queried just about any westerner about whether or not God loves them, you'd probably get something along the lines of, "Of course God loves me, I'm a good person," with considerable assurance.

These two young men and the young lady that I spoke to maintained a definitive and blasé confidence in their inherent goodness. Most would not consider themselves the potter or even the potter's wheel but would maintain that the value of the shape attained is as dependent upon the intrinsic value of the clay as it is the skill of the potter or even the quality of the wheel, a notion soundly refuted by Scripture. (Romans 9)

My one-year-old son aptly refutes the notion of intrinsic goodness. Little, pudgy, and snugly—the ladies at church absolutely cannot put him down and until two weeks ago, he ate it up. He smiled and flirted with them, almost intentionally placing them into a cuteness fervor. Now though, he hits them.

Ask any parent and they'll confirm that every child automatically knows how to hit, how to be selfish, how to be mean. You have to teach goodness and righteousness. As my son is going through his separation anxiety phase, he lashes out at whoever tries to hold him aside from me and Ami, and I mean everyone—male or female, young or old. We are currently teaching him to be sorry and show remorse. We never taught him to hit; we didn't need to.

Returning to the booth, I encountered a slightly different perspective. As I stood waiting, praying, observing the crowd, a man with his wife and son browsed the booth next to us. Sensing an opportunity, I moved closer.

"Sir, would you mind if took a few minutes of your time?"

The man looked at me, looked to our Christian materials on the table, and then proceeded to release some pent-up animosity. "No, that's fine. I have some questions for you anyway," was his emphatic response.

"Okay."

"Who created God? Hmmm? Who created God? And who did Cain marry and if you say his sister, the Bible forbids incest, so how is that right? How is that the deal?" was this man's angry introduction, lambasting me with his vitriol.

I backpedaled a bit internally, slightly taken aback and admittedly on my heels for a minute. I answered his questions as well as I could and as I recovered, we began a 30-minute conversation that actually ended up concluding cordially as he allowed me to pray for him and his family.

His eyes though. On our initial engagement, his steely blue eyes were harsh and penetrating, unwavering. As I prayed and presented the Gospel, he conceded that he was not a good person. He was a horrible person. He liked to drink, fight, and steal. He had been in jail and he didn't mind going back, and he didn't care who didn't like it. He wasn't concerned about God or hell, and if they sent him to hell they'd probably send him back, such was his badness. After all, he would likely end up running the place.

Now, as much as the confession of badness does not currently prevail in our culture, the other aspect, for those who concede badness, a proud humility in admitting badness has become prevalent. It's plastered on t-shirts and hats. "Heaven doesn't want me, Hell's afraid I'll take over." I am so bad that the devil could hardly contend with me, a seriously laughable notion. What the confessor of this particular attitude really says is that he does not seriously consider the notion of a creator.

"What about your son," I asked. "Would he say you are bad? Would you want him to see you as good?" At that, his eyes began to soften a bit. By the time I finished presenting the Gospel message, his eyes glazed with tears. I could almost see the Holy Spirit convicting his heart. As I said, we finished our encounter cordially enough and he agreed to allow me to pray for him and his family before bidding him farewell.

As postmodern man recoils at the thought of objectivity, the Bible frequently shocks with absolutes, consistently cutting across the grain of what most consider reasonable and rational. A century ago, the Bible shocked with the premise that God could in fact love a filthy sinner such as yourself. Today, the Bible shocks with the news that you are in fact a filthy sinner.

Men unavoidably inherited this condition as,

> *Therefore, just as sin came into the world through one man, and death through sin, and so death spread to all men because all sinned.*
> *Romans 5:12*

As the federal head, the representative for man, Adam's sin spread to all men, every single one. No man is immune to the ravages of sin. Sin has blackened every man's heart. Solomon writes,

> *...the hearts of the children of man are full of evil,...*
> *Ecclesiastes 9:3*

Jeremiah affirms,

*The heart is deceitful above all things,
and desperately sick;
who can understand it?*
Jeremiah 17:9

Men possess a sickness in their hearts that leads to folly. You've heard it advised, "Just follow your heart," an unbiblical and dangerous prospect. My friend who returned from combat in love with an Air Force captain followed his heart and in doing so, destroyed the lives of his wife and young daughter, tearing asunder a young family, crushing and shattering dreams and lives. He was just following his heart.

I've witnessed countless additional incidents of men "following their hearts" and normally also their genitalia as they walk away from wives, children, and families, killing hope. Much grief might be avoided if men would realize the truth of the natural wickedness residing within their own hearts, the total depravity of our condition.

Piper defines total depravity as,

> Man's natural condition apart from any grace exerted by God to restrain or transform man.[77]

This depravity, the depravity of men serves not only itself, but renders a man precarious and entirely hapless before a holy and righteous God.

The Hands of an Angry God

As godliness runs antecedent to all things worldly, Scripture frequently reminds us that God sees things differently than we do.

For the LORD sees not as man sees: man looks on the outward appearance, but the LORD looks on the heart.
1 Samuel 16:7

"I'm a good person. It's not like I kill people," some maintain. Others

---

[77]John Piper, "Total Depravity," https://www.monergism.com/thethreshold/articles/piper/depravity.html accessed on 7 Dec 2016.

chant, "Good without God," a popular contemporary atheistic slogan. You don't have to be a Christian to be good. Believers do not have a corner on the market for goodness. Now, we concede that many who are not of God accomplish much "good" in this world, or at least what we will call good.

We must make two points regarding these notions:

1. We must define exactly what we mean by "good".

To decide what is good, we necessarily must answer the question, "What or whom does it serve?" Who defines it as good? On what basis do you declare it good? Does "good" action or conduct necessarily imply "good" consistency or "good" substance?

Extremities may clarify the issue. Adolf Hitler returned home from Vienna after his mother was diagnosed with breast cancer. After a risky and drastic treatment, she was moved into the kitchen of Hitler's apartment where he tended to her, kept her warm, and did the household chores, anguishing in her affliction until her death after a short time.

Dr. Edward Bloch reported, as he settled the medical bills with a young Adolf Hitler that, "He had never seen anyone so overcome with grief," as the young Fuhrer was at the death of his mother. He also reported that Hitler expressed appreciation stating, "I shall be grateful to you forever."[78]

Adolf Hitler loved his mother and we may dub his care for her as "good" and even label his interaction with Dr. Bloch and his grief as "good". What of the clay? What intrinsic value did this particular lump of clay possess?

James writes,

> *Every good gift and every perfect gift is from above, coming down from the Father of lights...*
> *James 1:17*

Every man maintains a capacity for a certain level of "goodness". Total depravity does not dictate that men are as bad as they can be *all the time*. No, it implies that sin corrupted every aspect of man, even that which might appear "good", which yields to our second point,

2. As something may yet appear "good", sin taints even the

---

[78] "The Rise of Adolf Hitler," *The History Place*, accessed April 18, 2016, http://www.historyplace.com/worldwar2/riseofhitler/mother.htm.

noblest of intentions.

A particular account highlights this point. As the church service concluded and Charles Spurgeon visited with the congregation an older lady approached him and said, "You know, that was the best sermon I've ever heard preached on that topic."

Spurgeon reportedly responded, "You know, the devil just whispered that very thing to me."

When I preach the good news of the Gospel of Jesus Christ I truly seek to preach for the sake of Christ and Him crucified, for the love of the lost that they might know Him. I truly preach out of the desire for the reconciliation of the lost with the Master. There remains an ever-present temptation, a tendency to desire the approval of man.

The sermon is the same. The words do not change. It is my heart that so easily changes, always tending to default to self and sin. As I stand following a sermon, a part of me desperately desires the affections of men, and if I'm not intentional I might even wonder what is wrong if I don't receive much in the way of recognition. "Maybe it was a poor sermon?"

Sin has corrupted the human heart to the point whereby nothing good and pure originates from within. The human heart is deceitful and wicked and as such, nothing good, righteous, or holy springs forth from the hearts of men. One might accomplish "good" deeds, yet even these do not render the person as "good."

God looks to our hearts and He sees the blackness. He sees the pride and the lust, the sin and the self. God looks to our hearts and He sees that we do not pursue and love Him with the entirety of our affections as He commands. I could no more choose God and true righteousness than I could choose to change my gender. This condition demands justice and ultimately, judgment.

The Best Possible News

All religions outside of Christianity require supplicants to *do* something, to *earn* salvation. They never quite know if they've done enough. Many of the Muslims I've interacted with have a pall cast over their countenance, a darkness of eyes, a burdened posture even. They bear the weight of never knowing if they have done enough. Have they said the *Shahada*, paid enough Alms, prayed rightly, made the Hajj, and engaged in *Jihad*? Other religions follow suit.

Christianity says the opposite based upon the conditions of our hearts. God says we could never possibly do anything to earn the favor of God. Our being is absolutely corrupted starting with the wickedness in our hearts that manifests itself in wicked thoughts, words, and deeds. Now, we may possibly display a manner of outward righteousness, but it betrays the inner corruption that dooms us before a righteous and holy God.

It began in the Garden with his initial proclamation to Satan. (Genesis 3:15) God later made a promise, a New Covenant to fulfill and complete the Covenant of Grace that began with the promise to Abraham to bless the nations through him.

> *And I will give them one heart, and a new spirit I will put within them. I will remove the heart of stone from their flesh and give them a heart of flesh.*
> *Ezekiel 11:19*

> *For this is the covenant that I will make with the house of Israel after those days, declares the Lord: I will put my law within them, and I will write it on their hearts. And I will be their God, and they shall be my people.*
> *Jeremiah 31:33*

Just as God sees our hearts, He does His work on our hearts, on the hidden aspect of our being, changing us from the inside. Grudem defines regeneration as,

> A secret act of God in which he imparts new spiritual life to us.

This is sometimes referred to as being 'born again'. Scripture repeatedly affirms this critical and fundamental truth that the true work of God is in the hearts of His people.

> *They show that the work of the law is written on their hearts...*
> *Romans 2:15a*

> *When he turned his back to leave Samuel, God gave him another heart.*
> *1 Samuel 10:9a*

> *For with the heart one believes and is justified...*
> *Romans 10:10a*

Once God has given a man a new heart, once He has turned his heart of stone into a heart of flesh, the man now possesses the capacity to believe, to truly see his condition for what it is and to turn to Christ in repentance. With his natural heart, man rejects God. Regenerate man turns to Christ in belief that originates in the heart. Herein lies salvation, repentance, and belief originating from a changed heart.

Once God has changed a man's heart and the man responds in faith, the Holy Spirit indwells the man, empowering, equipping, guiding, and teaching.

and who has also put his seal on us and given us his Spirit in our hearts as a guarantee.
2 Corinthians 1:22

Jesus Himself likewise dwells in the heart of a believer.

so that Christ may dwell in your hearts through faith...
Ephesians 3:27

The real work of God is an inward work, a work of the heart. The things that we do matter infinitely less than what we are, the composition of our hearts. Here is the crux of man's dilemma before God, a dilemma to which He has provided the resolution in Christ.

A Heart for a Reason

Belief is not an end unto itself. God regenerates our hearts for purposes beyond salvation. Return to the New Covenant for the reason God regenerates,

...that they may walk in my statutes and keep my rules and obey them. And they shall be my people, and I will be their God.
Ezekiel 11:20

God saves for His glory *and* that He will have a people to call His own. He saves us into the Church, into the body of Christ. He grafts us into spiritual Israel, and as His people we live in fellowship with and obedience to the Lord. We seek to honor Him in holiness and purity, increasingly conforming to the image of Christ. We don't receive salvation and then cast aside all matters of faith, living as we did previously.

The converse proves decisively true, troubling in its implications. If you live exactly as you did before, if there is no change or no tangible indication in your life or heart that you are of God and that you have been saved, then what would lead you to believe you have been saved in the first place? Certainly, believers struggle with sin. In many ways, the life of a believer is very much a life of struggle. There must be some kind of change. If there is not, then that should be cause for some concern.

God has given us a new heart to love Him with and to love our neighbors as ourselves. True love, not the fickle romanticized pop culture notion of love, originates with God and springs forth through the regenerate hearts of men.

> *and hope does not put us to shame, because God's love has been poured into our hearts through the Holy Spirit who has been given to us.*
> *Romans 5:5*

> *If I speak in the tongues of men and of angels, but have not love, I am a noisy gong or a clanging symbol...If I give away all I have, and if I deliver up my body to be burned, but have not love, I gain nothing.*
> *1 Corinthians 13:1,3*

> *Love is patient and kind; love does not envy or boast; it is not arrogant or rude. It does not insist on its own way; it is not Irritable or resentful; it does not rejoice at wrongdoing, but rejoices with the truth. Love bears all things, believes all things, hopes all things, endures all things.*
> *1 Corinthians 13:4-7*

Questions concerning the condition of our hearts are the most important questions we might ever ask. Christ uniquely addresses our hearts. He reserves His most vicious condemnation for those who feign godliness while maintaining an internal wickedness. Christ condemns those who display righteousness though their hearts betray them to an all-seeing God.

Armed with a changed heart, I now possess the capacity and tendency to love with a godly love, with a changed heart. Worldly love says *I will love as long as I can receive. I will love someone or something as long as it can accomplish on my behalf, meet a need of mine, or do something for me.* In contrast, biblical love says *I love because Christ first loved me. I have decided to love no matter the outcome. I have decided to love, unconditionally, even those who can do absolutely nothing for me in return.* This is the very essence of biblical, godly love.

Curiously, the application of godly, biblical love proves decisive on the battlefield. Though you may find it unusual to speak of love in terms of war, perhaps sensing a conflict, I pray you will examine the idea along with me.

17. The Heart in Practice

Heart of a Man

In light of what the Spirit has revealed about the heart, how then should I act? How should I conduct myself during a time of war? How do we apply the knowledge of the Holy from the word of God to the individual actions of men confronted with conflict?

The plainest exhortation I can muster would be, "Grow up!" I don't know how to say it any clearer than this. In his exposition on love in 1 Corinthians 13, Paul provides a clear look at what true, godly love looks like. Interestingly, he concludes the section with the following admonition,

When I was a child, I spoke like a child, I thought like a child, I reasoned like a child. When I became a man, I gave up childish ways.
1 Corinthians 13:11

Paul encourages the Corinthians to love as God loves, to cast aside childish ways, and to take up the ways of a man, which is to love. To say it another way, the subjugation of self in regard to another in love is the business of men.

Real men possessing strength and courage ought to be in the business of denying self on behalf of others out of love. Men should deny self on behalf of their wives, on behalf of their children, and others. This is the truest essence of love at its most foundational. Men deny self on behalf of Christ…just as He did for us.

It is the childish man who imposes demands on behalf of himself. It is the childish man, lacking strength and courage, who refuses to subordinate his own self, refusing to consider others more important than himself. In other words, act like men. Quit worrying about yourself and love and serve others at all times, but especially during times of war.

The Warrior Way

The Warrior Spirit, often spoken of in military circles, becomes quite easy to codify and understand in light of the Gospel of Jesus Christ. Jesus said,

> *Greater love hath no man than this, that a man lay down his life for his friends.*
> *John 15:13*[79]

Jesus teaches the Apostles that the exhibition of the greatest of love is that a man will sacrifice for another or sacrifice his very life. One verse prior, Jesus stipulates,

> *This is my commandment, that ye love one another, as I have loved you.*
> *John 15:12*

Here, Jesus spoke not in abstracts as He embodied the very idea of this form of love. He loved us all the way to the cross. Jesus tells us to love one another and how to trade our lives for each other, to sacrifice our very lives if necessary on behalf of one another.

This does not necessarily stipulate death, though it may, but it does imply subordination. My wants, my needs, my desires, my concerns are secondary to those of others in *every* single thing that I do—loving my wife, raising my children, or engaging in war.

It is no coincidence that the most heroic actions from battles fought normally involve a soldier literally laying down his life to save his friends. Some accounts involve miraculous assaults against the enemy, but more often than not even these involve the protection of friends and comrades, outnumbered and outgunned. Almost always, the honoree has subordinated himself to whatever extent is necessary, even to the extent of losing his life. This is the essence of strength and courage, the essence of love.

We can see that engaging in battle and fighting in war is a supreme act

---

[79]Some passages, including the entirety of Psalm 23, I cannot help but quote from the King James Version of the Bible though I understand the limitations of the *Textus Receptus*.

of love. Sun-Tzu and Clausewitz miss the mark. Corporate and national hatred may drive at some levels, but this is not the true *Tao* to which Sun-Tzu speaks. The true *Tao* is the way of the cross, the way of love.

On the surface, reconciling love with the violence of war is counter-intuitive like many other biblical concepts. To kill and fight on behalf of your brother is an extreme act of love. To lay down your life for your brother is the supreme act of love. Because God has taught me and tested me, I know that evil exists and cannot be satiated. He has equipped me with the ability to fight, and he has armed me with strength, courage, and a new heart with which to love.

Thus, in love I fight on behalf of my sons that they may not have to. I fight on behalf of my daughters that they may never know the oppression of wicked men. I fight on behalf of my wife that she may live safely, far removed from the ambitions of evil. I fight on behalf of all who are oppressed by the corrupted hearts of evil men, and I fight on behalf of my brothers, my fellow warriors, for the truest glory and the greatest honor. There is no greater love than this, I assure you.

An Encounter with the Bearded One

I recall with very little trouble the most intimidating man I ever met in my line of work, though I've met other fairly intimidating men.

On my first ever training mission with the customers we were hanging out in the landing zone, waiting for them to come and get a passenger briefing from our crew chiefs. After a few minutes, the boys pulled up and marshaled about 20 yards away from the helo, content to stand and talk amongst themselves. We waited…and we waited. Five minutes became ten and finally, at the prompting of my crew chiefs, I walked over.

"Hey Sergeant Major, my crew chief is ready to give you guys a passenger brief."

With a whirl, the humongous, bearded assaulter turned upon me with eyes blazing and pointed a finger, "Look we ain't privates in the Ranger battalion!"

My heart nearly froze and a panic-induced adrenaline rush surged through my veins when the man's demeanor instantly changed. He smiled and said, "Just jokin' dude, we'll be right over. We're just going over a few details. Thanks."

I prayed that my countenance had not betrayed my heart. So I laughed it off, thanked him, and strolled back over to the crew chiefs where I relayed nonchalantly, "They'll be right over". I fought alongside this particular sergeant major for many years to come and I can say he was one of the most intimidating men I ever met.

Another time, I commanded under an officer who was one of the most intelligent men I've ever met. He was always a step or two ahead of everyone, including me, and I almost always walked away from our conversations kicking myself, "Why didn't I think of that?!" This man's cognition and grasp of next level and higher order concepts truly made him an intimidating presence.

I served underneath a different officer once who was so rigid and stoic in his adherence to basic Army standards, unrelenting in his pursuit of excellence in even the most benign aspects of military existence that he intimidated everyone that worked for him. Public beratings were not beyond his capacity, and I spent most of the year that I worked for him trying to avoid his ire and in actuality, his presence. He was an extremely intimidating man.

None of these men held a candle to the Bearded One.

The Bearded One resided in the lore of the community. As part of our training program, the unit sent me to SERE-C at Camp Mackall, North Carolina. SERE stands for Survival, Escape, Resistance, and Evasion with the C denoting that it was high-risk SERE, reserved for those who had a higher chance of becoming isolated behind enemy lines, thus presenting an elevated level of difficulty…and I had to face the Bearded One.

At first, I brushed off the accounts figuring that the Warrant Officers were just trying to scare us with stories of the Bearded One, kind of like the boogie man. Yet, the more I heard, the more I realized that this was real. The stories of the Bearded One at SERE were legendary and all that encountered him professed a deep-seated fear of him. I had to come to grips with the fact that I would soon stand before this man.

The course has changed in the nearly two decades since I went, but for several weeks, they taught rudimentary survival skills and how to conduct ourselves should we become captured. The course culminated with a week-long practical exercise. Inserted in small teams around the North Carolina countryside, we had to maneuver several miles each day through the thickest brush I've seen while evading capture. The only food we ate was what we could scavenge, which wasn't much. After several days, everyone was utterly exhausted, unbelievably hungry, and then we were captured.

A staged but dramatic event, the capture shocks your already demolished system—I lost 24 pounds during this final phase of the course. After a rigorous battlefield interrogation, they move you to the POW camp where you spend the next number of days. They do not handle you gently.

My first interrogation did not go that well, at least I don't think it did. Upon entry into the camp, they stand you before one of the cadre, kind of like a registration.

"Name!" He barks.

I mumble my name.

"Social!" He barks, sharper this time.

Without hesitation, I launch into one of my not-so-well-honed resistance techniques and mutter something about not remembering. The man freezes and looks up at me, his icy blue eyes piercing mine, penetrating.

"This man can see into my very soul," I remember thinking.

Possessing a slightly eastern European look, maybe Slavic, he got up from behind his desk and glided toward me and before I knew what happened, BAM! He slapped me across the face like I had never been slapped before. My ears rang in astonishment. The sheer reality of, "This dude actually hit me," flooded my senses.

The rest of the interrogation went equally as well as did the rest of my time in the camp, but thank God I had already encountered the Bearded One. At one time he had served as a cadre in the camp, but had been subsequently moved into the classroom.

About a week earlier, our class sat in a room awaiting our instructor for the initial block concerning conduct during capture. The door kicked in, and like the wrath of God himself, in strode the Bearded One. I nearly soiled my pants. He was tall and stout possessing large and powerful hands. A fierce mane of white hair crowned his head and he sported a Viking-like white beard but his eyes…and his voice. His eyes blazed from his face like two windows into the depths of hell itself and at the sound of his voice, every spine in the room stiffened, every heart skipped a beat. His voice was sharp and fierce.

"Class leader!"

It was the Bearded One. This fierce giant berated us for over an hour; calling us every name in the book, threatening us, impugning us, skillfully turning us against one another, honing in on any perceived weakness or division, driving wedges of dissent and hatred. When he walked out, all of us uniformly exhaled and sank into our seats, utterly spent and emotionally drained.

After about ten minutes of uncertainty, he walked back in. Only this time, his demeanor was different. Out of character, but still as intense, he spoke to us about captivity and I remember to this day, over 15 years later, what he spoke about.

He spoke about love. This fierce warrior that I described—and I realize after describing him that his description ironically bears a striking resemblance to that of the glorified Christ in Revelation—spoke about love. He said that it is love that motivates the soldier. It is love that drives a soldier. It is love that causes a soldier to sacrifice for his brother.

"You have to love your brother," he exhorted us fiercely, urgently.

This man, the Bearded One, who just minutes before had exploited our inner hatred and sin to drive a wedge of division, now exhorted us to love. It was the last thing I had expected to hear, but it stuck. Years later, it still resonates.

I discovered, after the course, that this man was a Sunday school teacher in his spare time, of all things.

Hearts Turned Wrongly

Atrocity is tough to ponder. It becomes easier to reconcile once considered in terms of the heart. As much as the Warrior Spirit is a manifestation of the Holy Spirit, the spirit of Christ, the spirit of atrocity is a direct reflection of the inherent wickedness found in the natural hearts of all men, of sin. As claimed earlier, all men possess deep within the desire and capacity to rape and murder, an unsettling thought.

Every honorable warrior is not a supplicant to Christ. Surely the majority who wage war in an honorable fashion do not follow Christ. Along with that not all soldiers who commit atrocity number among the heathen. Surely believers have engaged in atrocity, either actively or passively.

However, in any situation the wickedness of the human heart drives atrocity. Mitigating factors serve to assuage sinful actions. The sheer nature of depravity assuages sinful activity in war. As unregenerate men exist in a condition of total depravity, this doesn't mean they are always and consistently as evil as they could be. This only means that every aspect of their existence is tainted by the evil of sin. Unregenerate men exhibit a capacity for goodness or even morality, especially in regard to external actions.

Other factors check sinful behavior during the conduct of war. The common grace of the Gospel restrains sin even among those who do not profess Christ. Induced morality through upbringing and training restrains sin and produces a level of externally moral behavior. Societal norms and pressures drive actions, checking sin as do group dynamics and other social constructs.

Though the vast majority of the American military denies Christ, our professional creeds and mantras drip with Judeo-Christian heritage and values. The Army Values and the Warrior Ethos could have been grafted directly from the pages of the Bible.

However, issues arise when the societal and cultural norms drive and contribute to sin which, when organized by structuring forces such as military organizations, leads directly to atrocity. As an example, orthodox Islamic views concerning the subhuman status of infidels underwrites their corresponding maltreatment of prisoners and non-combatants. Religious norms like these directly oppose standards of biblical rightness and drive atrocity. Consider the suicide bombing of infidel civilians, a perfectly legitimate act in the eyes of any honest student of Islam.

These same things may also motivate believers to participate in atrocity and sin, if only by passively condoning. As much as any civilian believer struggles with sin, believers in the ranks likewise struggle. The terms of the struggle for the believing soldier just happen to potentially be more severe than for his civilian brother.

Was there not a single believer amongst the ranks of the Imperial Army at Nanking? Did no professing Christians know of the German atrocities during the Holocaust? Detractors love to reference the Crusades, though it proves a dishonest intellectual argument for the sake of undermining the Church. However, we cannot help but acknowledge the likelihood that believing Christians from Europe committed atrocities in the name of Christ, the difference being that no Christian will defend atrocity on any terms, even in the self-defense of an entire continent against corporate aggression.

Systemization and incorporation serve to amplify both goodness and sin as displayed in the singular actions of individual soldiers. Only the Gospel manifest in the hearts of men defeats sin and confronts atrocity.

The Switch Revisited

The fighting man, the believer, should seek to abolish the switch. There should be no switch. If you believe in Jesus Christ and you conduct yourself in a dishonorable manner that must be kept secret from your wife or your children or your friends, then perhaps you should reconsider how you conduct yourself in the performance of your duties.

Killing itself, though difficult, holds no shame. As we've seen, to fight and kill can be a great act of love and devotion when applied rightly. Do you engage in it rightly?

Let us revisit my "Muslim-killing" co-worker. As much as our actions were indistinguishable aside from our specific crew duties and occupational specialties, I will tell you that I seek to emulate God in that, "I have no pleasure in the death of the wicked." (Ez. 33:11) I will kill as called and able, in accordance with my duties with the utmost zeal and tenacity, but always with a heavy heart and a simultaneously internal reluctance.

I understand sovereignty and providence. I understand that God already ordained every action. I realize that my trigger pull is a slave to God and at the same time I'll offer up prayers for my enemies.

And I'll not have a switch. I'll proudly tell my sons about evil manifest, that God has called men to battle such evil, to stand firm against the wicked schemes of the devil. I'll speak in age appropriate terms. I'll refrain from relaying specifics concerning headshots versus body shots, the specific effects of different munitions. I won't tell them about the assaulter who tired of hearing the moans of the mortally wounded jihadi and shot him in the face. I won't tell them specifics about the atrocities committed by our enemies, that they like to saw people's heads off with a knife vice lopping them off quickly with a sword. I won't tell them that sometimes women and children die, even at the hand of those who seek to conduct war with honor.

With the clearest of hearts and with the firmest of convictions, I'll tell them of the greatest love I've seen, that of men laying down their lives for their friends. Do you exist in a dichotomy? Must you become someone else to perform your duties? If so, I'll suggest that you tread on dangerous ground.

Mother's Day, 2006 - Hearts on Display

A few poignant moments govern our lives and define our existence, and God offers a limited number of overt and obvious opportunities. I recall a handful of definitive moments that dramatically changed the trajectory of my own life. In July 2000, my young ex-girlfriend offered to walk away forever, knowing I had not intended to be a father at this point in my life. A year later, I chased this same young lady down Nashville's 2nd Avenue in the rain to ask her to be my bride. In May of 2007, I went forward before our church and declared God's call on my life to preach His word. The Grand Weaver, ever crafting our fate into reality, shapes all things into conformance with His will and good pleasure.

I often wonder about the moment of decision of this specific day, in this particular battle.

I didn't know Matt Worrell that well but we traveled together on this rotation, his last. En route, our C-5 stopped in Spain where we delayed 18 hours while the flight crew rested. Consequently, we rested, ate, worked out, and just kind of hung around. At one point, a bottle made the rounds and I vividly remember Matt commenting with a smile that he needed a toddy. As confident of a young officer as there was, he had a lovely young wife and two young sons who were truly his pride and joy. A Texas A&M graduate, Matt embodied Americana, patriotism, and all that we could be.

Jamie was different, a grizzled war veteran with over 25 years of service. He reminded me of so many other heroes I've known—quiet, unassuming, a family man. Having served in the unit for many years, his family was accustomed to his coming and going, yet it could never be easy for his wife and four daughters. I imagine they were looking with anticipation toward the end of his military service.

At some point during the battle, the gunship team was confronted with a decision. As high-tech as our military is, as much as our forces wield the latest and greatest equipment, sometimes the stuff just doesn't work. This is especially the case with communications. It always fails at the wrong time, right when you need to pass some critical piece of information or to hear an urgent order. Communications were an absolute nightmare on this day.

The commander, a friend of mine, and the rest of the headquarters were busy sorting things out. By now, word had spread throughout the command that there was a major engagement shaping up, and I can only imagine the 'help' my friend received. Assets were diverted, other forces marshaled, orders and suggestions given. The team had communications issues with the headquarters. They could hear their friends in trouble. They could hear their friends engaged in a fight for their lives. The launch order became confused. Launch or don't launch? Launch to where, to do what? To the laager site? To the target? In all this uncertainty, they launched…into the heart of darkness.

The sacred bond between the ground force and the attack pilots who protect them from the heavens runs deep, strengthened by years of sacrifice, watered by the blood of patriots and martyrs from times past. Matt, Jamie and the team stormed into the fray with the full certitude and authority of

God in heaven prepared to deliver judgment and death upon the heads of wicked men who sought the lives of their brothers-in-arms. Into the hornet's nest they blazed, into the killing fields, and within a pass or two over the objective the enemy blew them out of the sky with a heavy machine gun.

Major Matt Worrell and CW5 Jamie Weeks gave their lives for their friends, that they might live. I have no idea of their spiritual status. I only know what I could see, and that is that they fought with full conviction, exhibiting the love of Jesus Christ in the hardest and harshest of times. I will never forget their sacrifice and even more, their eagerness to sacrifice. They did not hesitate. They did not falter. They did not question.

These men sought to fight and even to kill, not out of hatred or anger, but solely out of the love that manifest itself from the very heart of Christ. Greater love hath no man than this, but that a man lay down his life for his friends. I am honored to have known these bravest of warriors and consider it a true privilege to know of this sacrifice.

Scarcely three weeks later, the task force killed AMZ in a kinetic strike just north of Baghdad.

18. Heart of a Leader

The Last Supper

Luke records at the Last Supper that,

A dispute also arose among them, as to which of them was to be regarded as the greatest.
Luke 22:24

The Apostles who spent the previous three years literally following Jesus, living with Him, eating with Him, and ministering alongside Him argued at the Last Supper over who among them was the greatest. Interestingly, this is not the first occurrence of such a dispute.

Jesus initiated the Passover meal with the Apostles by saying,

I have earnestly desired to eat this Passover with you before I suffer. For I tell you I will not eat it until it is fulfilled in the kingdom of God.
Luke 22:15-16

Jesus tells them this is His last meal with them, His literal last supper before He suffers at Calvary. They will not see Him again until He institutes the Kingdom of God. This is a serious and somber time and yet, they argue over who is the greatest.

From this, Jesus rebukes them,

The kings of the Gentiles exercise lordship over them, and those in authority over them are called benefactors. But not so with you. Rather, let the greatest among you become as the youngest, and the leader as one who serves.
Luke 22:25-26

After the rebuke, John records that,

> *Jesus...rose from supper. He laid aside his outer garments, and taking a towel, tied it around his waist. Then he poured water into a basin and began to wash the disciples' feet and to wipe them with the towel that was wrapped around him.*
> John 13:3-5

Follow along with what happened. At the most serious and somber of times, the Apostles argue over who is the greatest. Jesus rebukes them reminding them as He has before, that Gentiles exercise lordship, displaying authority, but the leader is "as one who serves" (Luke22:26). Then He displays exactly what He means. The Son of God—God incarnate, the image of the invisible God, the Firstborn over all creation—ties a towel around His waist and washes His disciples' feet. This is perhaps the most humble act of service in history aside from His passion on the cross.

On Leadership and the Bible

Per the Army, "Leadership is paramount to our profession" and true to form, the Army teaches leadership and leadership theory at all levels from basic training all the way to advanced schooling for senior officers and non-commissioned officers.[80] Army Doctrine Publication (ADP) 6-22 and its associated reference publication ADRP 6-22 teach the Army way defining leadership as,

> The process of influencing people by providing purpose, direction and motivation to accomplish the mission and improve the organization.[81]

It makes sense. Historically, the Army has asked soldiers to do difficult things, primarily to subordinate self on behalf of the greater good, sometimes to the point of death. Armies throughout history wrestled with this problem set. How should one go about motivating a soldier in this manner? For the United States Army, it's all about leadership. The leader becomes the critical component in generating purpose and motivation in soldiers, inspiring them to accomplish the impossible.

The United States even created institutions of higher learning designed

---

[80] ADP 6-22, Foreword.

[81] ADP 6-22, iii.

specifically to provide leaders for the military. I remember from my earliest days as a scared new cadet at West Point, hearing about leadership, studying leadership, memorizing quotes and sayings of famous leaders.

Though it has changed a bit over the years, I can still recognize the Academy's mission,

> To educate, train, and inspire the Corps of Cadets so that each graduate is a commissioned <u>leader of character</u> committed to the values of Duty, Honor, Country and prepared for a career of professional excellence and service to the Nation as an officer in the United States Army …

West Point's purpose:

> To provide the nation with <u>leaders of character</u> who serve the common defense.

A leader of character is an important person, invaluable to the common defense and service to the nation. Extrapolating, a leader of character would best provide the adequate purpose, direction, and motivation for the soldiers of the Army.

Virtually, the entirety of the Army's Professional Military Education system revolves around leadership. At some point, the Army leader makes a transition from direct leader to organizational leader. As a direct leader, usually at platoon or company level, you lead alongside the men, accomplishing the mission. "Follow me!" is your mantra.

The organizational leader, on the other hand, usually at the field grade level and up, leads organizations as opposed to individual groups of soldiers. Many successful company grade leaders fail to adequately transition to the field grade level, never grasping the intricacies of organizational leadership. The transition requires a definitive cognitive leap. No longer can the leader think in linear terms. He must be able to see outside of his current paradigm, unbound by anything but the limits of his own thinking.

Things Generals Say…

Everyone wants a piece of you at the pre-command course. Designed to prepare you for the rigors of battalion or brigade command, every officer preparing for command attends at least one of these courses. I attended the Army course at Fort Leavenworth and the aviation course at Fort Rucker.

Senior leaders (generals) flock to these courses to address those selected to command, and it makes sense. Insulated from the Army on the streets by numerous layers of command and bureaucracy, this provides the senior leaders (generals) with a chance to influence those who will have their hands directly on soldiers. This is an opportunity for them to impart their vision and guidance, and I was all in.

I was excited to hear from the senior leadership of the Army. I wanted to know what the vision was, what was next, where we were going, what we were going to do. I longed to drink the Kool-Aid. I was prepared to buy into what they were selling as, having been stuck in the trenches for the last ten years or so, I had no idea of what the grander vision might be.

The first encounter left me sorely disappointed. One of the most senior leaders in all of the military spoke to this room full of colonels, lieutenant colonels, and sergeants major via video teleconference and I remember getting up from my chair afterward thinking, "That was it?" Where was the vision? Where was the higher call? Where was our overarching aim?

This was his opportunity to influence the leaders of the Army who would directly impact the Army writ large and I couldn't have told you what we discussed. We spoke in vague generalities about 'getting back to our garrison roots' (it seemed the war was winding down at that point in time). We spoke at great length about sexual assault in the military, what a colossal problem it had become, and his fear that commanders might lose UCMJ authority to the civilians—other than that though, nothing. It was devoid of inspiration. I left the briefing deflated.

A retired general spoke to us at Fort Rucker. This man was one of the most successful officers ever from my particular occupational specialty, having risen to the highest ranks. He was polished, motivated, and intense as he communicated to us about the privilege of command. One thing he said stuck with me. He told us that while in command, we should have no other priorities, no other outside detractors, including our families or personal lives. We owed it to our units to make them our absolutely number one priority above all other things while in command.

I hesitated a bit and considered this thought. Do I really need to subordinate my wife beneath my command? Is this a necessary thing? Did I need to subordinate everything else to effectively prioritize my job in command? I agreed that I owed my soldiers my absolute best, but I wasn't sure that was mutually exclusive with other 'detractors' such as family. Yet, this guy was a general for a reason and I but a lowly lieutenant colonel.

The last speaker I remember was an ex-brigade commander, a colonel.

I remember that he was humble and quiet, speaking in hushed tones with a self-deprecating sense of humor. He spoke about a number of different things and then informed us that he would never prioritize the Army over his wife, and if that meant that he didn't obtain a certain level or position, then he was okay with that. I remember thinking instantly,

I'd follow this man through the very gates of hell.

The Army has since promoted this man to general, a decision with which I am extremely pleased.

The Army has always generated intense leaders and intensity is necessary. I want to be surrounded by zealots aggressively seeking the destruction of our enemies. I don't want a bunch of slackers defending my freedom and family from destruction at the hand of our enemies. At times though, it is those who sought advancement who progressed. At times, the Army promoted those who sought promotion the most, some who entirely served themselves.

Christ-like Leadership

Though volumes have been written concerning leadership and also concerning the leadership lessons of Christ, I'd like to consider three aspects of His character that translate directly to effective military leadership. I have become, over the last 20 plus years, a student of servant-leadership. I believe leaders exist to serve those they lead. Servant-leadership best accomplishes the purposes of leadership as stated in ADP 6-22, the accomplishment of the mission and the improvement of the organization.

Jesus washed his disciple's feet but more than that, He taught. We'll consider three separate aspects we might glean from this.

1. Humility.

Consider the great humility of Jesus and contrast that with the actions of certain leaders. Maybe you know them. In the military we exalt leaders and leadership to the point whereby we place leaders on a pedestal, officially and unofficially. At some point we lose accountability and eventually when the leader falls, as he's bound to do since he is a fallible man, we wonder, "How could this happen?"

In November 2012, it surfaced that General David Petraeus engaged in an extra-marital affair with his biographer, Paula Broadwell, prompting his resignation as the Director of the Central Intelligence Agency. It reminded me of a discussion I participated in several years prior.

In 2007-2008, I attended the Army's Command and General Staff College at Fort Leavenworth, Kansas, a school designed to prepare Army officers to serve at the field grade level. During one particular session, as my staff group discussed leadership, General Petraeus came up, as he previously commanded Fort Leavenworth. At Leavenworth, General Petraeus rewrote Army counter-insurgency doctrine, and the Army subsequently hand-selected him to implement that doctrine in Iraq the following year as the war had been going poorly for some time. Under Petraeus' leadership, coupled with President Bush's troop surge and the Sunni Awakening in Anbar Province, the tide of the war turned and General Petraeus' star continued its rise. Many whispered about political aspirations. He truly seemed to possess unlimited potential.

During our discussion, as several of my fellow officers had previously worked for Petraeus, we discussed in detail his methods and his leadership. Ironically, our teacher at some point asked, "What is the chink in this guy's armor?" He possessed no apparent weakness from what anyone could tell.

As you gain rank in the Army, people tend to want to do things for you. Some of it is ethical and some of it is not so much. What happened with General Petraeus? How did he become unaccountable? Where were his peers or contemporaries to ask him about spending so much time with a younger female who was not his spouse? Why did someone not say anything and save his career and the life and career of Ms. Broadwell before it was too late?

Because the Army relies so heavily on leadership and exalts leaders, we build a personality cult around those who rise to the highest level. They become nearly untouchable, insulated from all things including accountability. The vast majority of senior leaders who fail in the Army fail due to moral or ethical reasons. They've proven their professional competence. That is not in question. When you remove accountability from fallible men, the resultant fall is unsurprising, especially if they are men of means.

I have no idea if General Petraeus was a humble man or not but as the system exalts leaders, they must strive for humility. Look again to Christ.

As the eternal Son of God, the Lion of the Tribe of Judah, God in the flesh, He bared Himself and knelt, taking the feet of his disciples, those who studied from Him, and he washed them. He did this amazing act as they argued about greatness and who embodied greatness. Jesus says, here is greatness, and He gets on His knees and washes their feet. What an amazing display, counterintuitive to the mind of a natural man.

One may never truly serve as a leader without humility, and as Scripture reminds us, "Pride goes before destruction." (Pr. 16:18) To serve our men as Christ served, leaders must pursue humility. They must resist the temptation to be exalted and to allow others to exalt them. They must realize that they exist to serve and that God calls them to serve, especially when leading.

2. Stewardship.

Christ took complete ownership and responsibility for the Disciples and indeed, for all that the Father gave Him. He led them, guided them, and taught them in all aspects of their life. He took this mission seriously and consider that this band of 11 young men literally changed the world by, within a generation, proliferating the Gospel to all the nations.

Our standing in Christ drives our positioning before God. As we repent and believe, God adopts us into His family, as sons.

> *But when the fullness of time had come, God sent forth his Son...to redeem those who were under the law, so that we might receive adoption as sons. And because you are sons, God has sent the Spirit of his Son into our hearts, crying, "Abba! Father!" So you are no longer a slave, but a son, and if a son, then an heir through God.*
> *Galatians 4:4-7*

What an amazing thing. As adopted sons of God, I have full birthright. I am a fellow heir of all that God promises and most of all, I have a heavenly Father to whom I may now cry, "Abba! Father!" I am no longer a slave but a son. As a son of God, I have a much different standing before God than previously. My adoption governs all of my interaction with the Father.

The most effective leaders 'adopt' their soldiers. They love their soldiers and lead them as if they were their own sons and daughters. This love motivates the necessary heart for stewardship.

As a leader in the military, I have stewardship over the equipment that the nation has entrusted to me. I have stewardship over the lives of the young men and women that I have been assigned. I take responsibility for them and all that they do or fail to do. I take seriously the responsibility to train them, to prepare them, and to employ them. Never will I overlook my responsibility to the nation, to the parents of these sons and daughters. It is my commitment, my charge, to care for them as if they are my own.

This, by definition, is stewardship, caring for something that doesn't belong to me as if it were my own. I seek to be a good steward of that to which the nation has entrusted me. I seek to honor that commitment by giving my all to the soldiers under my leadership. As God has done for me, I do for them. Like Christ, I can do no less.

3. Service.

Lastly, in humility, I steward the most cherished assets of this nation, its sons and daughters, by serving them. Look again to Christ and the example He set as He served His disciples by the simple, menial task of washing their feet, a pointed rebuke to their argument concerning worldly greatness.

Previously, He said to them,

> *If anyone would be first, he must be last of all and servant of all.*
> Mark 9:35

The true leader, those who would be first in heaven, seeks to serve. They do not seek exaltation or position or authority or rank. They serve those they lead. Now, this service can take many different forms and may include things like discipline. What really matters though is positioning. Where does the leader rank himself alongside those that he leads? Does he truly see them as more important than himself? Does he truly seek to serve them, to subordinate himself, his own desires, needs, and welfare? Does the leader never ask of his soldiers that which he himself is not willing to do?

These three qualities—humility, stewardship, and service—best exemplify the leadership of Jesus which translates very effectively to military service.

Having had the privilege and opportunity to command on a number of different occasions, I've witnessed how soldiers respond to this type of leadership. Soldiers inherently understand when a leader seeks his own good. No matter how competent and professional that leader may be, the soldiers under his command will never willingly follow. They may follow solely based upon his professional competence, yet lack the inspiration to accomplish beyond the minimal.

The words of Paul exhort the believer but ring decisively true for the leader of men,

> *Do nothing from selfish ambition or conceit, but in humility*
> *count others more significant than yourselves.*
> Philippians 2:3

My most recent commander embodied this notion by serving in humility and never asking of his soldiers that which he would not do himself. His concern and affection for his soldiers emanated from everything that he did and it's not as if he didn't have high standards. He possessed the highest of standards coupled with the moral clarity to look a man in the eye and tell him that he wasn't meeting the standard, that he should find another occupation. His soldiers sensed all of this about him and they served him with a mutual affection. I would likewise follow this particular officer through the very gates of hell.

Inspiration

Men want to be inspired as they want to be led. Men desire to serve a higher call. Only the higher call drives men to subordinate themselves on behalf of a greater good. Men hunger for leaders to deliver this to them, the higher call.

I had the privilege of attending a recent address of a senior ground force commander to his soldiers. These soldiers had accomplished what few other ground forces in the world could have over the course of several months. They executed three extremely high-risk missions, successfully prosecuted the objective, and brought everyone home though a few had been wounded. These men sat before their leader, thirsty for the higher call, the bigger picture, the greater good, and they inquired about it. They stood ready to be inspired. Most of all, they wanted to know that their commander *believed*.

The commander chose to talk about administrative friction and camp conditions. Now, these are important things and they must be addressed. At some point though, all men want to be inspired, and what is most inspirational is when their leaders themselves are inspired. I don't know if this commander is inspired or not, but he did not communicate it to his men during this particular encounter.

Driven men generate ardent followers. Have you noticed that those who commit themselves to a higher call seem to produce committed followers? Very few men will fight to the death for a disinterested or professionally motivated leader. Men will slay themselves for a committed and inspired leader who has imputed that commitment and inspiration to them. They will accomplish the impossible.

Shortly after the attacks of September 11th, I sat in a hangar and listened to our commander speak. As we prepared to launch the initial wave of retaliation against our enemies he stood before us. He spoke at length about the enemy, about evil. He spoke about the privilege of being able to do something about the attacks and not just talk about it. He spoke at length about his young daughter. This rugged and gruff commander welled up a bit as he spoke of protecting his young daughter from evil men and doing all that he personally could to ensure that the world she inherits is safe and good.

He believed. He was inspired. He saw a higher call and the men responded. Just a few weeks later, they initiated the opening strikes against the Taliban and accomplished several missions that had never before been attempted or even imagined, truly amazing things.

Leaders have to buy in and believe themselves, live the higher call. When inspiration is absent, soldiers will follow a competent leader to a point, and perhaps your men will never be tested to push past that which many consider reasonable. However, the rigors of war might push you to call upon them, to ask of them that which most would consider unreasonable and they will want to know. Do you believe yourself?

As leaders, we must look to the source of our own inspiration if we are to ever inspire others.

A Last Thing

Thinking about leadership and thinking back to the Bearded One, I cannot help but recall his exhortation, "You have to love." You have to love your fellow soldier and as a leader, you have to love those that you lead. God calls us to love Him with all of our being and to love our neighbors as ourselves. As a leader, our neighbor definitely includes those that we lead.

The biblical leader loves his soldiers and serves them with such tenacity and zealotry that they likewise serve him with a mutual love and affection. You can tell a unit where the soldiers truly possess an unwavering love for their leaders.

Is it surprising to consider love in the context of the brutality of warfare, the harshest of tasks, the conduct of war? I don't know what else, other than love, could have driven MAJ Matt Worrell and CW5 Jamie Weeks to lay down their lives for their friends.

19. Consecrated Hearts

Let us return to Joshua, Canaan, and Israel and consider Israel's failure to conquer the Promised Land. As we talked about sovereignty and providence, we noticed that in some regard Israel failed because God ordained it to happen in His secret, unrevealed will. In the spirit of concurrence, let us examine Israel's actions and responsibility as we seek a secondary cause that may yield a manner of insight. Conjecture becomes necessary at some point but conjecture rightly made does not diminish the value of lessons learned.

Recall that Joshua marshaled his forces east of the Jordan River prior to the invasion and deployed spies into the land. As the spies returned and issued the report, Joshua and his forces prepared to ford the Jordan, initiating the invasion. He spoke to them saying,

> *Consecrate yourselves, for tomorrow the Lord will do wonders among you.*
> Joshua 3:5

He ordered his men to consecrate themselves the night before the invasion. The word consecrate, from the Hebrew *qadash*, means to set apart, to sanctify, to prepare. Joshua orders his soldiers to prepare their hearts for the work to come. As we've seen, war is a decidedly difficult venture. The taking of human life, the destruction of the *Imago Dei*, requires strength and courage. Joshua tells his men to be ready, to prepare their hearts, and to summon their strength and courage for this holy but difficult task to which they have been called.

They perform magnificently. The siege of Jericho happens exactly as the LORD commanded and the results are more than they could have hoped for. Things begin to fall apart almost immediately. Interestingly, never again does Joshua order his men to consecrate themselves, to prepare their hearts, at least as it's recorded in Scripture.

The way of the cross is one of struggle and more than that, the way of daily struggle. Jesus proclaims,

If anyone would come after me, let him deny himself and take up his cross daily and follow me.
Luke 9:23

Several aspects of this command bear consideration, but most relevant to this discussion is the modifier 'daily'. It is a daily consideration, a daily struggle. A Christ follower must take up his cross every single day and sometimes more frequently than that. Jesus understood the implications of David's words in Psalm 51:3 that his sin was ever before him. Every day, sin waits and its desire is for you, to consume you. Satan prowls like a lion seeking any he can devour. Every day, the ability to fall defines our struggle.

This is why a believer must take up his cross, consecrate himself, and prepare his heart every single day. Most assuredly, when he lets his guard down, the enemy will be right there waiting. Did Israel continue to consecrate themselves, to prepare their hearts or did they grow cold and comfortable in their commitment? The difficulty in the task at hand never diminished. In all reality, things got tougher as Canaan progressively fell. If anything, their task required ever more deliberate consecration, a steadfastness of heart never found in the strength of men.

As we see in returning to Judges 3, God desires to *test* and to *teach* through the conduct of war, and this is one of the lessons He desires you to know, that of consecration.

A Heart Tested

I wonder what went through Warrant Officer Hugh Thompson's mind as he rolled out of bed on March 16, 1968 and prepared for duty. Did he open a Bible? Did he pray? Did he consecrate himself for the testing yet to come?

A number of hours later, as he first maneuvered his OH-23 Raven over Task Force Barker's advance against the villages of My Lai, he and his crew noticed that something seemed amiss. Despite the absence of enemy fire, Thompson noted, "Everywhere we'd look, we'd see bodies."[82] He and his crew began to suspect that Americans were murdering civilians. As they

---

[82] "The Heroes of My Lai," accessed October 14, 2016, http://law2.umkc.edu/faculty/projects/ftrials/mylai/myl_hero.html.

hovered over a pile of bodies, they noticed movement, a little girl still alive. At that point, CPT Medina, the company commander, approached and kicked the girl over, gunning her down right in front of the startled crew.

Thompson and his two crew members became instantly aware that they were privy to a great tragedy. The unexpected circumstances confronted them, demanding a decision. They could not remain neutral, in the shadows. They had to decide; they could act or they could continue reconnaissance and turn a blind eye to the killing spree that raged beneath them.

Several hours later, LTC Frank Barker called an end to the search and destroy operation but not before upward of 500 civilians—elderly, women, and children—had been murdered in cold blood by soldiers from C Company under CPT Medina and LT Calley. It could have been much worse.

Warrant Officer Thompson and his crew did what they could to stop the bloodshed. At one point, he deplaned and confronted LT Calley on the ground. He evacuated civilians, placing his aircraft between them and advancing soldiers. He ordered his gunners to engage any American soldier killing civilians. He made frantic radio calls to higher headquarters alerting them that something was very wrong.

Thompson was tested that day, confronted with a great evil being perpetrated by his very brothers-in-arms and without hesitation, he stood in the gap, placing himself between the helpless and the danger that stalked them and for that he is a hero of the highest order. Amazingly, Thompson was vilified in his own country as the military attempted to cover up the massacre, even threatening him with court martial. Only later was he exonerated as the truth about My Lai was exposed.

Had he been ready, with a consecrated heart? Did he awaken that day knowing he'd be tested in the most impossible way?

Closing Hearts

God has called you to a most difficult task, the waging of war, for a very specific purpose. As you consider the task at hand, I'd ask that you consider the condition of that which governs you: your heart. Is your heart hardened to the Creator? Do you still resist the irresistible? Does the blackness of your heart betray outward manifestations of righteousness and goodness?

Would you submit? Strength and courage, necessary for the mission, are at your hand by the shed blood of a Savior who laid down His life out of an unwavering love for His people. Consecrate yourself that the Lord might do wonders among you.

Section 4:
War and the Soul

Chapter 20

Sunday, May 8th, Iraq

It's been four days. The first few days in theater are always a blur, getting a hand-off with the guy you're replacing, getting on cycle, getting up to speed on current ops, basically getting yourself sorted out. The guy I replaced left yesterday so I have the CON, no problem, no excuses.

I'm pretty much on cycle. I slept as much as I could en route and Benadryl helped out the last two nights.

Speaking of the sovereignty of God and His plans for our lives—I'm rooming with the acting CSM, a 1SG from the rear. He's a good dude that I've been around before but don't remember too well. Last night, as I sat down to read for a bit (I'm studying Habakkuk on this trip), he started asking pointed, open-ended questions concerning God and the Bible. It became apparent rather quickly that he and I rooming together was no accident. Some of his questions concerned the sovereignty of God and the seeming disparity between God of the Old Testament and God of the New Testament, something which I've been studying quite profusely for this very work, so I felt moderately equipped to speak the issue with him.

We had a great conversation lasting about an hour and, as he will be here for another month, I anticipate more. I pray that the Lord will speak through me into his life. It is yet amazing when I notice the hand of God, especially where I haven't expected to see Him. There He is, always working. I wish I could discern His hand more readily.

I am honored to work alongside the ground force once more. Not having served in the community in a number of years, I'd nearly forgotten what an earnest and committed group of men they are. I am always humbled to serve alongside them and am in utter awe of their courage on the battlefield and their blasé commitment to excellence.

A chaplain arrived today and announced that he is holding a service which I am excited about. Also, one of the ground force, huge dude with a beard (go figure), has a tattoo on his shin that says, 'Jesus Saves'. I am excited to interact with him at some point.

As dark of a place as the Army is, I've experienced some incredibly spiritual times downrange. My wife, of course, first pointed out to me the vast spiritual need the military has for the leadership of godly men. I'd been feeling the call to preach since shortly after my conversion, but I ignored it. I wrestled with God. Surely I was not called to this. I was an Army officer, a fighter not a preacher. What did this call mean? Did it mean I had to hang everything up and move on, go to a land as yet known in the tradition of Abraham? I languished for months.

Ami, sensing my frustration, pointed out what a great mission field the Army was. I am reminded of the words of Paul in 1 Corinthians 7:20, "Each one should remain in the condition in which he was called." Now, if God called me to go somewhere, I must obey. He had issued me a call, and I was already where I needed to be.

The Army is full of young men who don't know the Lord, young men who struggle with life. These young men, many of whom never had a father, need godly men to show them the way, to show them Christ. I've discovered that the Christian faith is so unlike anything they've seen that they respond greatly to Christ, to godly leadership.

When I assumed command three years ago, I had a series of meetings that afternoon with all of my leaders where I announced my goals and objectives as well as the fact that I was a man of faith and viewed all things through the lens of the Bible and that was just how it was. I cannot tell you how much their reaction surprised me. I literally had a line outside my office door of curious soldiers wanting to know what I meant.

When I speak as a commander, I frequently invoke Scripture and the Lord. Aside from one complaint, I have had nothing but positive feedback. I took my cues from a great chaplain I had who always invoked the name of Jesus in public prayer. He said to me once that he was accountable to the Lord above men and he wanted to error on God's side. I've taken that to heart though I've also taken care not to alienate those who don't believe as I do.

I believe that soldiers respond to sincerity and they respond to the notion of a higher call. The idea of good versus evil, godliness versus wickedness, resonates in the hearts of men and they respond.

I deployed with a Ranger battalion once and the chaplain stood before them to speak. He didn't tell them that they were going to do their duty, that they were deploying to support their country, that they were doing what they should do. Instead, he informed them that they stood as only they could, to confront evil itself and evil men who would seek to introduce their evil and wicked ways to our country, to our families, to suppress and subdue all that we hold sacred; that resonated. I walked away from that invocation ready to deliver as much death and destruction upon the heads of the enemy as I possibly could. The Rangers seemed to feel the same way.

I started preaching downrange. I kept ending up on small outposts with no chaplain, and at one point a group of Rangers had a small worship service and I volunteered to preach. I preached the rest of that rotation. Fast forward several years and I had made it a habit of deploying, linking up with other believers, and setting up Bible studies and worship services. It got to be where I absolutely relished this time and we developed some intensely tight fellowships to the point where I almost, not actually, but almost hated to go home and see our fellowship dissolve. The culminating tour was my first trip to Kandahar whereby we had a large group of about 20 dudes from several different groups in our task force and just had an amazing time of fellowship.

The ground force commander was a great believer and he and I became good friends. I cherish those memories of time spent serving the Lord in the darkest places. And oh yeah, after preaching to the ground force I can safely say that there is probably not an audience alive that could intimidate me!

(later)

I am so overjoyed to have discovered some other believers and plugged into a fellowship. God is so very good and faithful.

I lashed up with the huge bearded dude with the "Jesus Saves" tattoo. The chaplain sent an invite to his evening service. I debated about going and then figured, 'why not?', so I went. Though it was a small crowd of four men, there sat the dude with the tattoo. The service started a little awkwardly but once the young Air Force chaplain got going, he delivered a good message—more of a discussion, actually—from John 21, which the tattoo dude was deeply involved in.

After we concluded in prayer, I stayed for a few minutes to chat with the tat dude and learned that his name was Josh. As he had already been here for two months, he was in desperate need of some fellowship and study. We agreed to get a study/fellowship going, and I told him that I knew of a few brothers in my group that would love to be a part of it.

I am excited for this fellowship, just what I needed to not just get through these next months, but to prosper spiritually. There is something about the bond between men risking their lives in combat together that develops exceedingly strong spiritual bonds. I am praying for this time.

20. The Soul and the Law

As God teaches through the conduct of war, I've sought to impress upon you that all things possess a spiritual component. In actuality, the spiritual component always surpasses the visible aspects of existence in primacy. This understanding is easy enough to forget in the day-to-day conduct of life and certainly in the preparation for and conduct of war.

It is spiritual turmoil, failure to reconcile spiritual truths with the reality of existence that haunts many a warrior, yielding the glut of mental and emotional affliction that overwhelms our systems today. Wrongful spirituality drives atrocity, as the unchecked wickedness of the human heart, either enslaved to a collective effort or acting unaccompanied, enables the vilest of endeavors. Spirituality resonates on the battlefield. It was spiritual apathy and spiritual disobedience that condemned Israel's invasion of Canaan as they faltered, first in spirit and then on the battlefield.

Spirituality matters. The metaphysical and the supernatural matter greatly. This is a difficult concept in the mind of a warrior, one who engages in the most physical and real of actions, the application of physical violence against an armed enemy. Allow me to assure you: your soul matters. The condition of your soul matters, a truth I pray you will come to embrace.

Generating Valor

Armies throughout the ages developed a number of mechanisms to enable killing, to drive its warriors to kill more efficiently. Tactics changed, often accompanied by or even driven by changes in technology and technological leaps. An army digging trenches in the manner of World War One—much as ISIS does today in defending its territorial gains in Iraq and Syria—would not stand a chance against an army skilled at AirLand battle, an army utilizing mobility and initiative to out-maneuver static defenses. The 1st Armored Division would slice through ISIS lines in a matter of hours were they committed appropriately.

Militaries have increasingly sought to increase the gap, the stand-off that is the distance between the soldier and his intended target. A soldier further from his target is safer and also, more likely to engage. Today, surprisingly young soldiers slaughter the enemy in droves via a video screen. Drone pilots and drone warfare have changed the nature of conflict. It's not personal. Would any of these young soldiers drive a bayonet into the heart of his enemy or even shoot an enemy soldier while looking him in the eyes? Distance breeds effectiveness with respect to killing.

Not every effort addresses technology or tactics. Grossman discusses operant conditioning that desensitizes soldiers to the act of killing. This is why all soldiers of modern armies shoot at silhouettes vice the bullseye targets that were once so prevalent. Nations dehumanize enemy nations and enemy people groups to enable killing, a practice that continues today. There is no longer an official program, but it continues in an informal fashion. I can kill a 'haji' or a 'raghead' much easier than a father, son, or husband.

Lastly, militaries throughout the ages sought to foment a warrior spirit in the hearts of soldiers and leaders. From ancient Sparta to the 101st Airborne Division, nations have pumped up their soldiers and their bravado, knowing that at some point they may ask the impossible of them as at Thermopylae or Bastogne.

Armies are typically composed of predominantly young men, already possessing some natural tendencies that set them aside and render them uniquely appropriate for military service. Almost all violent crime is committed by young men as young men maintain a certain tendency to view physical aggression as a valid course of action. Young men possess a certain spirit that, when harnessed correctly and rightly motivated, yields excellence on the battlefield. Armies seek to direct this spirit, to harness the violent tendencies of young men, to shape them into a useful force.

Consider an airborne unit. Typically, not just anyone can join an airborne unit. You must become airborne qualified at some point by passing airborne school. Some airborne units maintain a rite of passage or a tryout of some kind. In some way, you must first establish your mettle prior to being considered for inclusion in the ranks. You must account for yourself and show yourself worthy of inclusion. Once a member, you are introduced to the higher standards, the demands for excellence. Though just a new guy, you quickly buy into the mantra of the airborne, forsaking all others as inferior.

Airborne ideology is infused into your heart and mind at all levels. You are made to feel special, elite, set apart, different and most of all, better. The feeling is quickly welcomed as most young men seek inclusion and to be a part of something bigger than themselves, something better. Young men want to belong.

My friend who served in the 82nd Airborne Division described for me a post-wide run whereby if any fell out, they were ordered to remove their unit shirts and run bare-chested in the wood line lest they be seen and bring discredit to their unit. I'm not sure what they did for females. Some airborne units have their own creeds that soldiers recite ad infinitum. All of this exists for a singular purpose, to enable.

I remember well my very first jump at airborne school from a C-141. Preparations start at a ridiculously early hour. This is it, the day you have been waiting for. You don your chute and wait to be JMPI'd—have your equipment checked out by the jump master—and then you wait…and then you wait some more, for hours. Finally, you line up and shuffle to the aircraft. They pack you in like sardines, your belly full of butterflies.

The aircraft takes off and flies for what seems like forever and then finally, the door opens. The door of the aircraft opens in flight and you are firmly confronted with the reality that you are about to jump from a perfectly good and functional aircraft! This is not a natural act. The jumpmaster runs through his shouted commands. You echo them. At some point, you stand up and hook up your static line. It's all scripted; it's all routine and rehearsed by this point. *Thirty seconds*! The roar of the jet engine overpowers your senses. Adrenaline surges through your veins. This is it!

The light turns green and the jumpmaster shouts, "Go!" The line bustles toward the door. You follow closely the man in front of you, not wanting to get left behind. It's a blur. There's the jumpmaster. You thrust your static line to him and out the door. The violent impact from the jet blast, the jerk and shutter of the opening shock and then…silence. You look up and behold the most glorious of sights, a fully round and open chute doing its job, delivering you safely to the ground, provided you keep your feet and knees together.

It took two weeks of training to get to this point but why? Literally, the entire act involved sitting there, getting rigged up, walking to the aircraft, following the man in front of you and then stepping out the door. Why would it take two entire weeks of training to get you to this point?

Because jumping out of a perfectly good aircraft is not a natural act. It is something that requires a steeliness of heart and nerves, a fire in the soul. I've always hated jumping, always been scared to death of it, but I've loved serving in an airborne unit because I love the fire in the hearts of an airborne soldier.

For the first two weeks at airborne school prior to jump week, they pump up your spirit. Yes, you learn mechanics and processes, practicing the few things you have to learn ad nauseum until you could execute a parachute landing fall (PLF) in your sleep. Mainly though, the cadre seeks to infuse the airborne spirit into the students.

That is the point. Build up your spirit to overcome the natural resistance to jumping from a perfectly good aircraft. You build it up through the repetitions of the body and the training of the mind. The black hats at airborne school really serve as mental coaches, forging the spirit of the airborne into the students. That is the entire point of the course.

It doesn't stop there. As much as young men continuously seek ways to test themselves, the military provides opportunity after opportunity to test yourself, to prove your mettle, to number yourself among the elite. The military esteems the soldier and the challenge, and they adorn those who successfully complete the next challenge with badges and titles and medals and even headgear. The maroon beret of the airborne is but a badge, a symbol that you are cut from a different cloth than those not similarly adorned. We exalt those who have been tested and found worthy.

All of this is by design, in reality, to overcome the ultimate thing that, as a creation of God yourself, you will resist the most: the taking of human life. It has proven remarkably effective as militaries have become experts in fomenting the warrior spirit. Couple this with innovation in tactics and technology and the dehumanization of the enemy and men have developed a great capacity in modern militaries to deliver death and destruction *en masse*.

In all of this I wonder, to what end? What is the price for such a spirit? What happens when a soldier, built into a hyper-actualized image of himself yet lacking a foundational spirituality, is confronted with the destruction of human life?

The Imminence of Confrontation

Inevitably, the warrior stands, naked and stripped of all pretense, defenseless with his soul laid bare. Confrontation looms. At some point, the soldier and his ethos run headlong into a *seemingly* oppositional construct. He may for some time bury the confrontation and its associated effects under a glut of bravado or merely allow the passage of time to dull any aches. At some point the bravado loses its efficacy and the years fail to soothe. All defenses eventually prove futile over time when confronted with the reality of death.

The actual grit of battle obscures honor and things considered honorable. If you've ever seen, in person, what modern munitions do to a frail human body, consider yourself blessed. I've never seen a dead body on the battlefield and thought of honor and God. Instead, men in violent death often assume the appearance of a caricature, a mockery of the previous glory of life and vitality. The stark nature of death and confrontation with death actualized, particularly death at your own hand or even the hand of the man to your left or right, assaults the senses and challenges previous conditioning.

In response, the battle-hardened warrior's heart might develop a great callous. The callous may conceal and even protect a bit, but it hinders reconciliation. Many a warrior considers this a perfectly acceptable condition content to exist in a continual state of subtle tension between anticipation and anxiety. Still, confrontation looms. Might I suggest that confrontation on favorable terms would be a proper and sought-after course?

The Soldier and the Schoolmaster

Let's return to Ron Kovic anguishing to his mother, "Thou shall not kill, mom. Though shall not kill!" His confrontation, though dramatized for effect, is not uncommon as all men, apart from a psychotic few, possess a God-given resistance to killing his fellow man. The requirement to induce men to kill opposes this natural resistance. What is a man to do at this point?

We'll seek to transform our minds via an understanding of the actuality of eternity and the soul, a necessary step for successful reconciliation. And so we'll ask, what of the law?

Ron Kovic's devout Catholic upbringing infused within him an appreciation of the Ten Commandments. This is what we must do, obey the Ten Commandments. To do otherwise is sin and places us out of the graces of God, in opposition to Him, which is where Kovic found himself in his heart and soul, hence his anguish. He anguished that his country called him to something that placed him in opposition to God as he saw it.

Kovic's character possessed a perverted understanding of the nature of things, of the relationship between man and God's law codified in the Ten Commandments. Granted, this is a fictional Hollywood account based upon a real person and a true story, but in the account Kovic lacks a true understanding of the nature of the law, something which men have been victim to since its inception.

The average citizen could no more name the Ten Commandments than define the actual purpose of the Ten Commandments. Most could regurgitate a couple and then tell you these are the things that God has told us to do perhaps to get to heaven. If confronted with the fact that they have broken the commandments, they would respond that God would forgive them. This train of thought, completely prevalent in contemporary western culture, is anathema to the word of God.

Most people consider the Ten Commandments an outdated standard bearing little relevance to real life. From the Bible, the law is an integral aspect of Christianity without which, true faith does not exist. A proper understanding of the law, of why God gave us the Ten Commandments, confronts numerous spiritually erroneous positions that afflict so many.

Let us first look to the words of Paul,

So then, the law was our guardian until Christ came, in order that we might be justified by faith.
Galatians 3:24

Here, he speaks to the law as a guardian, a *paidagōgos*, a servant whose job it was to take the children to school. In Roman culture, servants took the children to school and would often instruct. The NASB translates the word as *tutor*, the King James as *schoolmaster*. Each rendering provides some insight, but what exactly does Paul mean when referring to the Ten Commandments as a guardian, a tutor, or a schoolmaster.

Paul explains to the Romans,

...Yet if it had not been for the law, I would not have known sin. For I would not have known what it is to covet if the law had not said, "You shall not covet."
Romans 7:7

Paul says that he would not recognize sin apart from the law. The law is our schoolmaster, our tutor, and what does it teach us? It teaches us that we are sinners in need of a savior. When confronted with God's standard, the Ten Commandments, I am reminded that I lack the ability to obtain the good graces of God, that I could never meet His standard. God's standards, the Ten Commandments, prove impossibly high.

"But what if we keep most of them?" you may ask. Scripture clarifies this as well from James,

For whoever keeps the whole law but fails in one point has become accountable for all of it.
James 2:10

James tells us that if we break a single commandment, we stand guilty of breaking the entire law. If you have told a single lie, you are guilty. If you have stolen a single thing, you are guilty. Jesus ups the ante in the Sermon on the Mount, ascribing the law to the condition of our hearts. Thus, if we *look* at a woman with lust, we are guilty of adultery in our hearts. If we *hate* a man in our hearts, we are guilty of murder. The rich, young ruler asks of Jesus, "What must I do to inherit eternal life?" (Luke 18:18) and Jesus responds with "You know the commandments." (v.20) The young man responds with, "All these I have kept from my youth," (v.21) an obvious lie.

Jesus is not trying to save the young man with the law, but rather point out to him that he maintains no hope of saving himself. The law tills the rocky soil of the human heart and identifies within us our helplessness and our need for a Savior. The law softens hearts, breaks hearts, convicts our spirits, and turns us to Christ.

Here we begin to understand. Recall the paradigm that enslaves westerners whereby most assume God's good graces because they are such good people. Unsurprisingly, the proliferation of this attitude occurred simultaneously with a diminished prevalence of the Ten Commandments. What a simple yet effective tactic of the Devil. Relegate the law to obsolescence and undermine the Gospel at its core. On the other end of the spectrum, the Devil distorts our understanding of the purpose of the law.

Attempting to keep the law out of coercion or in an effort to obtain a level of righteousness is known as legalism, a fatal mistake made by men from the very beginning and a concept condemned harshly by Christ. As a matter of fact, Jesus reserved his harshest criticism and judgment for the outwardly righteous, those who appeared to keep the law to gain God's favor yet were wicked on the inside.

The order proves critical and decisive. I do not keep the law in order that I might become righteous before God. No. Rather, because I cannot keep the law, I understand that I need a Savior. And when God saves me through faith in Jesus, I then seek to keep the law out of commitment to Him, having been empowered by the Holy Spirit.

The Ten Commandments, for the believer, become a guide and a way to live the Christian life, but not a way to obtain salvation. He already granted that through faith in Christ. The law remains a schoolmaster, calling the believer to repentance whenever necessary, and guiding his daily life.

Legalism, a false understanding of the law, inevitably generates guilt. Guilt is a cognitive or an emotional experience that occurs when a person believes or realizes—accurately or not—that he has compromised or has violated a moral standard and bears significant responsibility for that violation.[83] How many warriors languish unnecessarily under the guilt of having killed? Yes, taking human life is difficult and hard, as it should be, but what of the commandment?

Exodus chapter 20 records the giving of the law. The sixth commandment generates the confusion in this regard. The King James renders the commandment as,

Thou shall not kill.
Exodus 20:13 (KJV)

Herein lies the confusion. Clearly God calls a man not to kill. Why would He not hold me in judgment if I do the very thing He has prohibited?

First, remember the purpose of the law. I am not saved, my salvation does not depend upon my adherence to the law. My salvation depends upon my faith. If one were to kill, no matter the reason or situation, he would not necessarily be condemned or rejected. Following salvation, God will call and empower the believer to comply with the law, generating another reasonable query concerning the actions of the believer in the taking of life.

Accurate wording is critical as the King James, unfortunately a popular and even exclusive translation among many bodies of believers, provides a faulty rendering of the Hebrew word *ratsach*. It is most accurately translated here as murder. Both the ESV and the NASB (along with the HCSB), the standards for modern biblical translations, translate the word as murder. The KJV does as well in the majority of its other uses in the Old Testament.

As Jesus addresses the issue in the Sermon on the Mount He uses the Greek word *phoneuō* which like its Old Testament counterpart, can be translated as either murder or kill. In Matthew 5:21, the KJV sticks to its rendering of 'kill' while the ESV and NASB maintain the translation of 'murder'. In Matthew 19:18, the King James deviates and translates it *murder* instead of kill which the ESV and NASB affirm. Matthew 5 and Matthew 19 share an identical context. The inconsistency of the King James Version undermines its veracity on this point.

When interpreted from Scripture, we must retain the translation of *murder* with respect to the 6th commandment. The most accurate

---

[83] "Guilt". *Encyclopedia of Psychology*. 2nd ed. Ed. Bonnie R. Strickland. (Gale Group, Inc., 2001), eNotes.com. 2006. 31 December 2007. accessed May 18, 2016, http://www.enotes.com/homework-help/psychology-what-guilt-what-stages-guilt-466309.

understanding of the sixth commandment is "You shall not murder." (Ex. 20:13)

The difference between murder and killing involves intent or malice. Do you kill for justice, for a just reason, or do you murder with evil intent? Justified killing, in line with Scripture, would include defending yourself or your family against aggression, the performance of civil duties such as that of a police officer, and the conduct of war. War does not unilaterally justify killing. We cannot say *carte blanche* that all wartime death is justified; think of My Lai or Nanking. We must apply what we have learned. What is the spirit behind the action, what is the heart and the motive?

The taking of human life demands special consideration but does not necessitate an absolute prohibition. Understanding this proves necessary in the reconciliation we seek along with the avoidance of misplaced guilt. As men live enslaved to legalism, the servitude of a warrior is an exceedingly tragic affair. We must cast aside these heretical notions and turn to the reality of God's word on the matter.

Chapter 21

21. The Soul of a Warrior

Consider if you will once more, the feminization of the contemporary church and indeed, the feminization of the west writ large to include this nation. Modern western society marginalizes, decries, misunderstands, and in some respects, even demonizes true manhood. The west has become a society devoid of any sense of real masculinity, authentic biblical masculinity.

Several competing forces pervert the collective masculinity of western man. Feminist elements seek to temper masculinity, to quell the brazen souls of men, to smooth them out, to soften the rough edges. Public schools, institutions of higher education, other civic institutions, and even governments contribute to this phenomenon whereby the ideal for man has become …a woman. Western culture seeks to neuter men at many different levels. We medicate young boys for acting like, well, boys, and shame and demonize all things of masculinity.

Men have acquiesced and abdicated their biblically mandated responsibilities to lead: in the family, then the church, and finally in defense of the nation. Contemporary western man is fully free to pursue whatever gratification of the flesh he desires as he is likely unencumbered by any of these responsibilities and nations are weaker because of this. Families absent male leadership are weaker. The Church, with diminishing male participation and leadership, is weaker. Western nations continue to crumble at the undermining of the foundations of the family and the Church, largely due to the absence and apathy of men.

Each passing generation perpetuates and exacerbates the issue as without godly men and fathers to teach them how to behave, young men fail to internalize concepts of biblical masculinity and manhood. They usually err in one of two ways. Some bend to the will of the effeminate masses and allow themselves to be emasculated and shaped into flaccid, feckless, and hollowed-out parodies. Still others resist in the opposite manner and oppress and exploit females as a show of strength and faux-masculinity. Either situation embodies the curse from the Garden. (Genesis 3:16)

We've seen that warfare is one of the toughest things, one of the most difficult to ponder, which is why the Bible repeatedly exhorts men to 'be strong, have courage,' to 'consecrate themselves'. Warfare is a decidedly masculine activity involving the harnessing of masculine traits—the desire to conquer, to achieve, to overcome. Audacity and boldness, traits dripping with masculinity, undergird decisive military operations. Few things are manlier than closing with and destroying the enemy in battle.

Once, strong men roamed free, unencumbered by the chains of feminist expectations: men of valor, men of courage, men who scoffed at death and the devil. Let us look now to such men and learn from those who have fought. Let us gaze upon the soul of a warrior and lay bare his heart that we might know where our own might yet stand.

A Man After God's Own Heart

David was a man after God's own heart.

Acts chapter 13 records that Paul and Barnabas arrived at Antioch in Pisidia, and they went to the synagogue on the Sabbath to hear the reading from the Law and the Prophets. The synagogue leaders invite them to share any words of encouragement. Paul stands and delivers an address, a lengthy sermon culminating in a presentation of Christ as the fulfillment of the Messianic intentions of the Law and the Prophets. At one point, he references David, essential to the lineage of Jesus, referring to him as a man after God's own heart. (Acts 13:22)

David was a man after God's own heart.

About 1000 years prior, in the Valley of Elah, a young David stood astride the prostrate body of the giant Goliath. With a grunt, he drew the fallen champion's sword, and with a thrust, finished him off, killing him. He then cut off his head and brandished it to the now emboldened Israelites. The Philistines, their champion dead, fled in terror. David humbly came before King Saul still carrying the head. Eventually, he brought it to Jerusalem—Goliath's head, that is. (1 Samuel 17)

David was a man after God's own heart and God was seeking such a man as him. (1 Samuel 13:14)

Exalted Warriors

The Bible commends Joshua—commander, warrior—and as Scripture records, "his fame was in all the land." (Joshua 6:27) Let us return to the invasion and the idea of God *testing* and *teaching* through warfare. In considering the warrior and his position before our Lord and by extension, the position of all warriors before the Lord, we will allow Scripture to instruct. In this, I assert that you will notice quite the opposite of what some might expect, and that is the exaltation of the honorable warrior.

David lived by the sword for the entirety of his life, and he surrounded himself with men who did the same. Scripture records a group, a special few, the *Gibboram* or David's mighty men. These were the elite, the most distinguished fighters of the day, known as the thirty (there were actually 37). Abishai, commander of the thirty, wielded his spear against 300 men and killed them all becoming the most honored of the group. (2 Samuel 23:18,19) Benaiah struck down two of Moab's best warriors, killed a lion, and later a 7 ½ foot tall Egyptian with his own spear. Renowned among the thirty, David placed him in charge of the royal bodyguards. (vs. 20-23) Scripture lists the others by name including Uriah the Hittite, a character we'll revisit shortly. (vs. 24-39)

Within the thirty was still another group, the three, the most renowned of all fighters. Josheb-basshebeth was chief of the three. He killed 800 in a single engagement. Eleazar stood by David and defied the Philistines even as the rest of the Israelites withdrew. He fought until weariness overtook him and "his hand clung to the sword." (v. 10) Thus the LORD honored him and delivered a great victory unto Israel that day. The last of the three, Shammah, likewise stood and fought the Philistines as all of Israel fled and in much the same way, the Lord delivered a great victory to His people. (2 Samuel 23:8-39)

At one point, the three came to David in the stronghold of the cave at Adullam. The Philistines were encamped in the Valley of Rephaim with their garrison at Bethlehem. David made an off-hand comment about desiring a drink from the well at Bethlehem. The three subsequently go to Bethlehem, break through the lines, draw water, and take it to David. David, surprised that they did such a thing, refuses to drink the water asking, "Shall I drink the blood of the men who went at the risk of their lives?" (2 Samuel 23:17) Such was their loyalty to their warrior-king, David.

These warriors from antiquity, the thirty and the three, fought with the full conviction and authority of the LORD our God. They fought fiercely with courage and valor. They did not whither in the face of the enemy, *even when those around them did.* They stood and they fought and God honored them and even saw fit to record their names for antiquity in His holy word. Recall that *all* Scripture is inspired by God and is profitable. The mention of these men of renown, these men of courage and valor, is entirely intentional. Could it be that men need to hear and learn of such things, lest we forget what God has called men to do?

The Lord sought men such as this, men such as the three and the thirty, men of courage and conviction, men such as David. He still seeks such men.

David the Warrior

The reign of David was born of the disobedience and weakness of Saul, his predecessor and Israel's first king. From presenting unauthorized sacrifices to ultimately failing to completely destroy the Amalekites in battle as God ordered, Saul's disobedience is consistent and his rule is likewise doomed. Samuel the prophet declares to him,

> *The LORD has sought out a man after his own heart...*
> *2 Samuel 13:14*

Saul was not that man, a man after the LORD's own heart. Later, after the failure with the Amalekites, Samuel declares,

> *Because you have rejected the word of the LORD, he has also rejected you from being king.*
> *1 Samuel 15:23*

At a later battle, the arrows of a Philistine archer found Saul on Mount Gilboa. As his armor-bearer refused to end his life, Saul slew himself with his own sword. (1 Chronicles 10:4b) His son Jonathan, along with his three other sons, also died at the hands of the Philistines that same day.

David grieved as did the entire nation yet, "all Israel gathered together to David at Hebron," (1 Chronicles 11:1) and anointed him king. This fulfilled what God declared many years before when Samuel anointed the humble shepherd David, the youngest of 8 sons. "This is he," the LORD declared to Samuel of David. (1 Samuel 16:12) The LORD had been seeking after such a man and had finally found him.

David's way was the way of the sword, from his earliest days to his death after 40 years of rule. As a young shepherd, he struck down lions and bears that threatened the flock. (1 Samuel 17:34) He slew the champion Goliath, brandishing his severed head for all to see. (1 Samuel 17) He fought the Philistines and was pursued by Saul, who was wary of being usurped by the upstart David. David became a vassal of the Philistines (1 Samuel 27) and ultimately defeated the Amalekites. After being declared king, he united the kingdom following the assassination of Saul's son Ish-bosheth. Later, he conquered Jerusalem and defeated the Philistines and eventually Ammon and Syria. Later in his rule, he faced two separate insurrections from his own sons, Absalom and Adonijah. David knew little of peace during his days, and after 40 years of rule, death drew near.

On his death bed, as was the custom in those days, David summoned his son Solomon. "Be strong, and show yourself a man," the dying warrior-king exhorted his son. (1 Kg. 2:2) Further, "keep the charge of the LORD your God, walking in his ways." (v. 3) One can envision this proud old warrior, gray and frail, yet still with a hardness pulling his young son close, struggling for breath, urging him to be strong, be a man, follow the LORD and also, put to death his enemies that remain. (vs. 5-9) In other words, be strong, be a man, follow the LORD…and fight.

David breathed his last and then "slept with his fathers." (1 Kings 2:10)

This was the man that the LORD had sought. This was a man after the LORD's own heart. This was the man that women celebrated singing,

Saul has struck down his thousands, and David his ten thousands.
1 Samuel 18:17

Of David, the LORD declared,

I have granted help to one who is mighty;
I have exalted one chosen from the people.
I have found David, my servant;
with my holy oil I have anointed him,
so that my hand shall be established with him;
my arm also shall strengthen him.
The enemy shall not outwit him;
the wicked shall not humble him.
I will crush his foes before him
and strike down those who hate him.
Psalm 89:19-23

The warrior-king David fought with the strong arm of the LORD by His hand and it is the LORD who brought him victory, who crushed his foes, and struck down those who hated him. David's pursuit of the way of the sword was a most holy pursuit. His was no tempered pursuit, no life of mediocrity and contentment. He fought and at death, was welcomed by the Lord into His presence. He was a man after God's own heart.

And so God finally appointed this wizened, battle-hardened warrior-king a time of rest "with his fathers." He had fought the good fight, finished the race. No more would he take up arms though I have to imagine, even in heaven, that David's hand still itches to wield the sword, his fingers still tremble just a bit when recalling the days of yore.

David the Sinner

David's transgression and subsequent fall are the stuff of legends. David, the warrior-king, at the zenith of his power falls for what else but a woman.

At the time, Scripture records that David had at least seven wives and numerous concubines. As the king, he possessed limitless opportunities to satiate any sexual urges. He possessed power and wealth above all others yet, he chose to plunder another man's fields. And not just any man, but a faithful servant and valiant warrior, Uriah the Hittite.

"In the spring of the year, the time when kings go to battle," David sent his commanders and army to fight the Ammonites and lay siege to Rabbah while he stayed behind in Jerusalem. (2 Samuel 11:1) The rest of chapter 11 records the sordid details of the encounter. One afternoon, as David strolls on the palace roof, he notices the exquisitely beautiful Bathsheba bathing. He sends for her, takes her, and she conceives.

David, now fully immersed in his sin, sends for her husband Uriah from the front and gives him leave in Jerusalem, hoping he will lay with Bathsheba and cover his sin. Uriah, an honorable warrior, refuses to go home and lay with his wife while his brothers are at the front fighting. "As you live, and as your soul lives, I will not do this thing," he declares before David. (v.11) David dines with him and gets him drunk, and still Uriah refuses to go to Bathsheba.

Finally, with no choice left, David returns Uriah to the front with orders for his own death. David directs Joab the commander to place Uriah at the front where the fighting is hardest that he might fall, which ironically, is probably where he would have been anyway. Joab obeys and Uriah falls in battle the very next day. After a period of mourning, David takes Bathsheba as his wife and she bears him a son, though Scripture records that, "the thing that David had done displeased the LORD." (v. 27)

Consider the blood on David's hands. He had been fighting for the better part of his life. He beheaded enemies and slew a vast multitude, but it is this betrayal that displeased the LORD. Before, he had fought with honor. Here he disgraced and dishonored himself before the LORD.

One may glean numerous lessons from this account. As we continuously consider the Lord's view of warfare, consider where David was, where he had chosen to be. It was the spring when kings went to war, yet he remained in Jerusalem and chose to send his army without him. Had David been where he should have been, the episode likely would not have happened. Scripture is silent on the reason why David was not at war. Was he detained by other administrative requirements? Had he lost heart? Had he become fatigued or even comfortable in his position? Clearly, his desire to avoid warfare displeased the Lord. He should have been fighting alongside his men.

The sexual sin is obvious so we won't tarry there, but consider another aspect: the betrayal. Consider that this warrior-king, a man of honor, broke bread with his brother-in-arms, sat across the table from him, and looked this man in the eye knowing that he had already betrayed him. Consider that he ended this man's life to cover his betrayal and to make matters worse, Uriah would have given his life to save David's such was his courage and commitment. The utter betrayal of the warrior code underscores the entirety of David's sin.

The prophet Nathan confronts and rebukes David in a dramatic fashion. (2 Samuel 12:1-15) David repents immediately and declares, "I have sinned against the LORD," to which Nathan responds, "The LORD has also put away your sin; you shall not die." (v. 13) Despite his great sin, the LORD spares David though not without consequences as He declares through Nathan that "the sword shall never depart from your house" (v.10) and that his and Bathsheba's child shall die, which comes to pass. David's actions, though forgiven, yield consequences as turmoil consumes the remainder of his days.

Yet, as horrific as his sin is, consider that David is still a man after God's own heart as he pens Psalm 51, one of the most heartfelt lamentations of remorse and repentance in all of the Bible. David begs, "Have mercy on me, O God," (v.1) and "Wash me thoroughly from my iniquity" and "cleanse me from my sin!" (v.2) David mourns that, "my sin is ever before me" (v.3) and most of all that,

Against you, you only, have I sinned and done what is evil in your sight,
Psalm 51:4

Though David sinned against Uriah and Bathsheba, betraying them, he ultimately sinned against the Lord. He goes on to beg, "Purge me with hyssop" (v.7), "Create in me a new heart, O God" (v.10), and "Restore to me the joy of your salvation." (v. 12) David's anguish is true, his confession heartfelt.

Many misunderstand the idea of confession, that you merely tell the Lord about your sin. "Lord, I have sinned, here it is, forgive me." True confession, closely associated with repentance, is an affair of the heart. Confession has its roots in Greek language meaning "same speech". When one truly confesses, they say the same thing about their sin that God does. They view it the same way, are disgusted by it. They hate it. Sorrow consumes them but it is a holy sorrow, not a worldly sorrow at having been caught in iniquity.

David's confession and repentance truly reflect his heart and rather than dwelling, even in this great Psalm of repentance, he moves on to appeals of restoration and trust. "O Lord, open my lips, and my mouth will declare your praise," he goes on to proclaim. (v.15) In true warrior fashion, he acknowledges responsibility for his actions, repents, and gets back to work loving God and serving him.

Consider that the Lord sought David years before with the full foreknowledge that David would one day fall. It's not as if the encounter surprised God. No, in some way this encounter served the purposes of God and worked together with other things for the good of those who love God, perhaps if only to serve as a lesson to speak to us from the word of God on this matter.

Notice Paul's language at Antioch of Pisidia. Notice that he did not reference David as a man after God's own heart who also committed a great sin. Paul knew of David's fall with Bathsheba yet like God, Paul saw him for who he was, a man after God's own heart.

David the Worshipper

One need only look to the Palms to see David's heart for worship. Worship is a shortened wording of "worth"-ship. To worship is to literally declare the worth of something, its value. David declares,

> *O LORD, our Lord, How majestic is your name in all the earth.*
> *Psalm 8:1*

I will be glad and exult in you;
Psalm 9:2

The LORD is my chosen portion and my cup;
Psalm 16:5

The heavens declare the glory of God, And the sky above
proclaims his handiwork.
Psalm 19:1

All the prosperous of the earth eat and worship; before him shall
bow all who go down to the dust,
Psalm 22:29

Of the 150 Psalms, David composed roughly half of them. In David we see a merging, a coalescing of the warrior heart and a heart of worship.

1 Samuel chapter 16 records that "the Spirit of the LORD departed from Saul, and a harmful spirit from the LORD tormented him." (v.14) His servants brought David to Saul and whenever the spirit was upon Saul, David would play the lyre. It says that "Saul was refreshed" and "the harmful spirit departed from him." (v.23) David, described by the servants as "a man of valor, a man of war," (v.18) played the lyre. Consider that the same hands and fingers that smote the enemies of the Lord held the lyre and played beautiful music unto the Lord. What a delightful contrast we see in the ugliness of battle and the beauty of music in praise of God on high.

Years later, as David had the ark moved to Jerusalem, to the City of David, a great procession formed. This was a time of great joy and celebration, the culmination of an epoch. There was shouting, praise, and music, "and David danced before the LORD with all his might." (2 Samuel 6:14) David, warrior-king, man of valor, man of war, abandoned any foolish pride and danced before the LORD with all his might. Michal, the daughter of Saul, came and rebuked him for disgracing himself in front of everyone. David responded by reminding her that he was chosen by God before drawing the line, "I will celebrate before the LORD," he proclaims. (2 Samuel 6:21)

David declaration reminds me of Job's declaration,

…choose this day whom you will serve…But as for me and my
house, we will serve the LORD.
Joshua 24:15

Joshua says you can worship who you want. You can worship the false gods before you or you can worship yourselves or things created by men, but we will worship and serve the LORD. David's declaration is similarly definitive.

He has drawn a line and declared, this is who I am. I will celebrate, I will worship. I will serve the LORD, our God. That is who I am whether you like it or not. That David's willingness and his desire to be seen in this regard would inspire all men to similar convictions, this is my prayer today.

A Warrior of Conviction

The earnestness of David's convictions stands in stark contrast to a watered-down culture of men who stand for nothing. David's frank declarations and honest pursuit of the Lord starkly opposes the moral and spiritual ambiguity of nations that have lost their way. Do any even ask anymore, "What is worthy?" Do any even inquire? David had no issue in that regard as he poured himself into the noblest of endeavors—the conduct of warfare and worship of the LORD. In all of this, David remained a man after God's own heart. Might I suggest, that you could also be a man such as David.

22. War and the Soul

A Good Death...

I'd like to die a good death.

What concerns us most when considering matters of the soul but death? Has there ever existed a singular phenomenon over which we held less control? It seems, on occasion, that the entirety of human existence revolves around the avoidance of death, the prolonging of life. Now, life is precious, an absolute gift from the Lord not to be squandered. Yet, consider the frantic efforts to prolong that which the Lord has already determined.

As Christ reminds us,

And which of you by being anxious can add a single hour to his span of life?
Matthew 6:27

The Psalmist likewise declares both the transience of life,

Behold, you have made my days a few handbreadths, and my lifetime is as nothing before you. Surely all mankind stands as a mere breath!
Psalm 39:5

and the predetermination of the Lord in establishing our lives,

The steps of a man are established by the Lord,
Psalm 37:23

God has ordered our steps, every one, and despite our best efforts to prolong life, our *appointed* time will come. Do not become confused. Allow me to reiterate the sacred nature of life. We should seek medical care to prolong life. We should take care of our bodies as the temple of God. We should not act rashly in preserving and taking life. We must forsake all concepts that diminish the value of life including abominations like assisted suicide.

What I speak of is the "cure at all cost" attitude that so pervades the medical *industry* in this age. As columnist Courtney Martin notes, we should "actually do less for the dying."[84] Dr. Allen Frances in the *Psychiatric Times* advocates that, "none of us can ever cheat death and instead should strive for a good death that occurs at the proper time and in the proper place."[85] He goes on to describe a typical hospital death as "lonely and fearful; tubes invade your body and monitors constantly beep and bing; the environment is frantic, noisy, indifferent, brightly lit, and sleepless. You die among strangers with little chance to say goodbye to loved ones."[86]

What fear might drive such indignity that occurs in preserving life? My uncle, a D-Day veteran from Normandy, went to the hospital with chest pain. A CT scan indicated a dark spot on his lungs that might have been cancerous. Even though he was in his 80's, the doctors recommended exploratory surgery. The spot turned out to be nothing, but he contracted an infection from the surgery Despite the hospital's best efforts, he died a couple of weeks later…in a hospital with tubes in his body, monitors beeping, a cold and impersonal environment if there was one.

What madness might overcome us as we break the bones of frail old ladies doing chest compressions to milk a few more days? What psychosis is this that we stick an old man full of needles, and tubes, and drugs to squeeze a few more days, maybe hours? To what end? Against the value of life, do we see absolutely no value in death, death with dignity, death with honor?

My wife worked as a nurse for many years and confronted death personally on more than a few occasions. She described for me essentially two types of death. The first die peaceably, in spite of whatever indignations are thrust upon them by the medical industry. Many go smiling, talking to people not present. They are more than ready to slip into the dark stillness

---

[84] Courtney E. Martin, "Zen and the Art of Dying Well," *The New York* Times, accessed May 29, 2016, http://opinionator.blogs.nytimes.com/ 2015/08/14/zen-and-the-art-of-dying-well/.

[85] Allen Frances, "Dying Well Means Dying at Home," *Psychiatric Times*, accessed May 29, 2016, http://www.psychiatrictimes.com/couch-crisis/dying-well-means-dying-home.

[86] Ibid.

of death. Others go fighting, scared, often crying. Many frantically resist, seeming to know and realize the imminence of an unpleasant fate.

I'd like to die a good death—maybe surrounded by family as they read the word of God to me or maybe on the battlefield, standing firm against the evil in this world. I know the choice is not mine.

Death is the historically universal scourge that confronts all men. Listen to the words of Paul quoting the prophet Hosea,

"O death, where is your victory?
O death, where is your sting?"
1 Corinthians 15:55

Death no longer has power; death no longer bears a sting. Though I cherish life, I've come to understand that there is no tragedy in dying well, a good death. Christ assures us of this.

An Inevitable Fate

Eternity confronts all men. Whether we like it or not, the Bible tells us that,

...we will all stand before the judgment seat of God.
Romans 14:10b

and,

...each of us will give an account of himself to God.
Romans 15:15

Let us put this in context. As a believer in Jesus Christ, I have the blessed assurance of knowing that Jesus, the only man to have lived a righteous and sinless life, imputed his righteousness to me and God declared me innocent. I have been justified, and when God looks upon me forever more He doesn't see Brad Smith, the great sinner. He sees the righteousness of Christ. Such as it is for all who believe on the Lord, Jesus.

However, I will give an account before the Lord. I will give an account for my idle words and deeds, opportunities and time wasted. And as a servant of the Lord, my sole desire on that day is to hear the beautiful voice of Jesus gently whisper, "Well done, good and faithful servant." (Matthew 25:21) Either way, I know that He will remember my sins no more. (Hebrews 8:12, 10:17, Isaiah 43:25) My justification and position is as secure as the reality of God Himself. What an astonishing notion, certitude.

Imagine never having to wonder.

The Reprobate, those who will not believe, face a harsher reality. They are left standing bare before the judgment seat of God. They do not have the hope of the imputed righteousness of Christ. They have not been justified, declared righteous and innocent before a holy and wrathful God. These are the children of wrath and their judgment is as certain as it is just. God's holiness demands that He hold them accountable for their sin.

On that day, all will stand before Him, the books will be opened and if anyone's name is not found written in the book of life, he will be thrown into the lake of fire. (Revelation 20:15)

It doesn't matter what you do, how hard you try, even how religious you are. When it comes to salvation, it doesn't matter how often you go to church, whether you were born into a Christian family, or raised as a Christian. One of the most troubling passages in all of Scripture comes from the Sermon on the Mount. Jesus says,

> *Not everyone who says to me, 'Lord, Lord,' will enter the kingdom of heaven, but the one who does the will of my Father who is in heaven. On that day many will say to me, 'Lord, Lord, did we not prophesy in your name, and cast out demons in your name, and do many mighty works in your name?' And then will I declare to them, 'I never knew you; depart from me, you workers of lawlessness.'*
> Matthew 7:21-23

On that day, the Day of Judgment, many will stand before God and declare the things they did in His name and these are, by all means, not small things. They prophesy, they cast out demons, and do mighty works in the name of Jesus, yet He does not know them and will cast them into the outer darkness where there is weeping and gnashing of teeth. "Depart from me, I never knew you," the most frightful declaration in the entirety of Scripture.

The implication is that many today believe that they possess a certain position before God based upon the things that they do. To carry it further and to its most troubling conclusion, many who occupy a church pew likewise believe they are in a certain position before God based upon the things they do, but they stand condemned because they do not know Him and more importantly, He does not know them.

I find this perhaps the most troubling consideration, that people will die believing in the security of their salvation only to be confronted with a far different reality. I can imagine the surprise in the hearts of men who find themselves shockingly condemned.

Things That Matter

I've always searched for significance, wanting to be a part of something bigger than myself. I believe this is what drew me to the Army, the opportunity to subordinate myself to a higher purpose, a great call. Yet, as much as I love the Army, it could never quite scratch the itch.

On my second assignment in Honduras, I was designated the unit's tax assistance officer and sent to Panama with a nasty old Air Force sergeant for two weeks to learn how to assist soldiers in preparing their taxes. I remember being very depressed at the notion of 1) spending two weeks in Panama with a chain-smoking crusty old Air Force E7, 2) actually learning how to do taxes, something which I really had no interest in, and 3) not flying. As a young lieutenant, I wanted to fly. I was a decent pilot and I desperately desired experience, closely associating my worth with my number of flight hours. I loved to fly.

I've dabbled with journals and diaries over the years, never maintaining anything consistently. However, I wrote vigorously during my two weeks in Panama, lamenting my condition, my search for relevance, my search for significance. I felt useless, unimportant. Depression coursed through my veins.

Years later, that struggle persisted. I read that men go through several distinct stages of development. In the first stage, men desperately seek independence, particularly in the latter teenage years. Men seek to detach from their parents, their caregivers, and establish themselves in the world. After that, men seek experience. They seek to build their street credit. At some point, men seek to set themselves apart from their peers. They have the experience, the street credit. Now, they want to distinguish themselves in some way.

Lastly, once they've obtained and arrived at their intended destination, they seek redefinition. They examine that which they've obtained and realize it's not what they thought it would be and so they seek after something else. This presents the classic mid-life crisis. Men change jobs, buy a sports car, or maybe discard a spouse for a younger version.

At the time of my conversion, I found myself in the third stage. I had walked out of my home at the age of 18 and never looked back. Headed for West Point, I told my parents that I wanted to fly up by myself, unaccompanied. I wanted to plant my flag in the world. They humored me but I didn't know, as they were traveling quite a bit at the time, that my mother demanded my father take her to the airport in Newburgh, New York. My mother literally watched me deplane and link up with my sponsor. Some flag planting! I didn't find out about that until years later.

However, I planted my flag and then sought experience. Upon graduation from West Point four years later, I sought assignments overseas in Korea and Honduras. I'd volunteered to extend in each location to obtain experience. I tried out for and was selected into special operations aviation where again, I sought experience.

September 11th happened and I desperately sought deployment. I remember a wise old CW5 giving me some advice after my platoon did not get picked to be a part of the initial invasion into Afghanistan and I raged against the machine, offended to the core of my soul that me and my men were not chosen. He advised me to be careful what I wished for. In hindsight, after 16 years and numerous deployments, his wisdom rings true, but I obtained what I sought: experience.

As I approached my promotion board for major I began to fantasize about promotion below-the-zone, ahead of my peers. I had a decent file, not too shabby.

Why not me?

What if they pick me?

Wouldn't that be something?

It was about that time that I was converted. There exists a phrase that I loathe. It goes something like this, "Jesus is a gentleman. He gently knocks on the door of your heart, but you have to answer and let Him in." It's normally based upon faulty exegesis of Revelation 3:20 and has led to this moderately biblical phenomenon of "inviting Jesus into your heart." You hear it during church invitations and altar calls. You hear it told to our children at vacation Bible school.

One reason I write this is to declare to you that the Lord Jesus Christ kicked in the door of my heart with the full authority of the God of the Universe. He never asked. He never knocked. I wouldn't have answered! The Lord breached the door of my heart in a most decisive and violent manner. What I seek to convey is that yes, I believed, but I really had no choice. I couldn't not believe. That's the thing, though I could never articulate it at that time.

I sought religion and I found Jesus. More appropriately, He found me. Around the time that I truly longed after significance in my job, I found true significance in an eternal call. I found significance in the Source of all things and that significance still governs my life to this day. I didn't get promoted below-the-zone.

Confession

It occurred to me that perhaps you have participated yourself in what you might perceive as atrocity. Maybe you've not killed, maybe you've murdered in the conduct of war. Perhaps you've fought with an impure heart. As much as I've sought to speak to killing in the context of war, we must address the opposite aspect, that of dishonorable killing or even atrocity.

I only know to proclaim the truth of Scripture concerning the matter. God could never forgive me—a common lament and misperception among men. Know that you do not suffer uniquely in this regard. Maybe your burden is unique but the fact of your burdening is not unique. Many countless souls toil under such a misguided perception.

Please allow me to reassure you of a final truth. The only sin that condemns in finality is the sin of disbelief. John declares,

> *If we confess our sins, he is faithful and just to forgive us our sins and to cleanse us from all unrighteousness.*
> *1 John 1:9*

We come to confession in one of two conditions. If you are not of Christ, if you have never been redeemed, then I promise you have more pressing concerns than even something as serious as atrocity committed. You exist unreconciled with God, destined for judgment…unless you ask for forgiveness of your sins, specifically the sin of unbelief. And so I exhort you with all urgency, get on your knees and repent. God is faithful and just to forgive you of your sins and cleanse you from all unrighteousness. (1 John 1:9)

Maybe you come before Christ as a believer. The words of John still hold true. Repent. If you've killed wrongly, participated in atrocity, or fought with impure motives, take them to God. Confess, and he is faithful and just to forgive you your sins and cleanse you of all unrighteousness.

What an amazing thing that no matter our condition, the answer remains consistent: confess and be saved or confess and be healed. Either way, we find the answers on our knees, in confession and repentance.

Scar Tissue

This work was born of affliction and love. First of all, I love God and the Lord, Jesus. I have a difficult time imagining, even remembering my life before Him, apart from Him. I remember my desperate search for significance, my failed attempts at satiating a deep need in my soul. My only regret in life is that it took me so long to discover Him though, in the spirit of this work, I acknowledge that it was He that found me just over a decade ago and that He left me in my affliction for all those years for a very specific reason that I've yet to discern.

As much as my love for God and all things pertaining to God has driven this work, I've labored on behalf of so many of my brothers-in-arms who do not know Him. I love soldiers. They are rough and often crude and, at the same time, completely hilarious. I remember with great fondness many of the antics I've been privileged to be a part of, things that the civilian will never quite understand.

I recall with great fondness my crew-chief 'Powder' who literally leaped from our hovering helicopter to the jungle hillside below at a brief query of the possibility. I remember looking out my chin bubble as he rolled several meters to a stop before issuing a thumb's up. I fondly recall the grouchiness of my very first 1SG who hated warrant officers with an absolute passion and merely tolerated lieutenants as best as he could. I had a commander once who loved to gamble on everything, and I do mean everything. I recall getting the worst butt-chewing of my career from an angry colonel and the various other wire-brushings I received over the years.

I remember the best stick of gum I ever chewed. We had been out marauding on a long day in central Afghanistan and were on the hour-long flight back to Kandahar. I put our maintenance pilot in the cockpit and jumped in the back, sitting next to some of the ground force, several of whom were fellow believers, our legs dangling from the door of the helicopter over the vast expanse of the Afghan countryside.

We were hot, tired, and spent, and one of the young assaulters tapped me on the shoulder. Turning to him, he smiled and offered up a stick of gum. I took it, and to this day I recall with amazement that it was the absolute best stick of gum I've ever chewed. We sat in the back of that helicopter as it sped over the painted desert sands and smacked our lips, relishing the burst of flavor that cut through the grit of our bodies and souls.

We chewed our gum in solidarity, extracting every drop of flavor and all was right with the world. No matter the war, no matter the life taken, the life given, all was well. I looked around the back of that helicopter and every single assaulter sat deep in thought, relishing the moment, chewing their gum, completely oblivious for that moment, to the plight of men, the suffering of the masses, the fight that tomorrow may yet hold. How could I ever explain a thing like this to anyone but a soldier?

I love soldiers and I see the affliction of soldiers. Young soldiers are the most overtly afflicted yet, I believe that many gray beards quietly bear the scars of years of persistent conflict, decades of warfare. Even the hardiest of souls, after spending years engaged in the conduct of war cannot help but bear some burdens.

Growing up, my friend's dad embodied the Vietnam veteran for me. He was a gruff and burly construction worker and I remember being quite intimidated by him. At some point, he came and spoke to my high school history class as a part of our study of the Vietnam War. He had been a door gunner in Vietnam and as he spoke of the conflict, of friends lost and lives taken, he teared up. This gruff and burly construction worker teared up for a minute, collected himself, and then finished talking.

I remember being shocked at the dichotomy of manliness and grief, but now I consider that this man, though still grieving, did not become consumed. He raised his family, loved his wife, and lived his life as well as he could. And apparently, when the grief came, he dealt with it and then got back to the business of life. He didn't stuff it down per se, he just refused to be defined by his scars.

The scars are real, I know. The scars bear consideration. This too, I know. I had another friend who had witnessed a singular atrocity, one that most would consider very minor, and he told me at one point of the scar that it left on his soul. Considering this, I've quietly thought about those who have scars on top of scars, those whose souls have been repetitively lashed by the scourge of trauma, guilt, and grief, those for whom there seems to be no relief, those who carry burdens to which there is no respite.

I can only offer you one thing, and that is the truth as I know it, that which I have sought to communicate to you. Know that many, out of the best of intentions, offer many different things by way of solutions but that there is one Healer and His name is Jesus. Have you considered this? Though He may yet heal the wounds of your soul, I assure you that when you reach for Him, you'll find two nail-scarred hands to cling to. He knows your affliction because He has been afflicted in much the same way, yet without sin.

Would you seek Him today? Would you turn to Him today?

Redemption

My salvation was born of the crucible of peace. Fogginess clouds my memories of those days. I recall events but they are diluted, as if I were in a dream, almost as if another were living out these events.

My wife and I made a New Year's resolution. Our oldest daughter started to drift just a bit and so we decided she needed church. We would take her to church and they would fix her there. She needed the moral teachings of the church! Yes, that was the ticket. Looking back, I laugh at the sheer blindness of my perspective. I was the one that needed what I could not fathom.

Our first church visit proved less than fruitful. Only about 15 folks graced the sanctuary that Sunday and they instantly recognized us as visitors and they singled us out. Strike one. I cherished anonymity. Then, a squirrely guy with a guitar sang a few songs about Jesus. No hymns. No organ. No choir. Strike two. Lastly, they didn't even have a pastor or a preacher or a reverend, no professional, just some dude called an elder who talked for a few minutes about some things, and not very well from what I can remember. Strike three. I remember walking out and saying to Ami, "I got absolutely nothing out of that." Funny how perspectives change.

Next, a friend of Ami's invited us to her church, a large Southern Baptist Church, and it was here that I recall hearing the Gospel of Jesus Christ for the very first time. Their pastor preached the Gospel with full authority, with all its teeth. I'd never heard anything like it, at least that I remember. I grew up going to church, but never recalled hearing these things. Perhaps they said them and I didn't listen, but I never remembered hearing the Gospel prior to this.

For the very first time, God confronted me with the reality of my sin, my separation from Him, and my need for a Savior. I learned of my inability to save myself of my utter depravity, and for the first time I heard of the atoning work of Jesus Christ on the cross as He died for all who would believe. It rocked my world.

For the weeks leading up to Easter, the pastor would conclude each service with an altar call, an invitation to come forward and be saved. Every week I literally resisted with all my strength. I clung to the pew with white knuckles. The Lord had changed my heart and I wanted to believe, but I clung to my flesh fiercely, to my old self for as long as I could until His grace overwhelmed me and I could resist no more.

We attended our church's Easter play. At the end, the pastor issued an invitation, that any who might believe on the Lord Jesus would be saved. I raised my hand and said "yes" to Jesus. I did not know that I was in effect signing the death certificate for Bradford Smith the great sinner, while simultaneously receiving the birth certificate of Bradford Smith the new creation in Christ, saint according to the grace of our Lord Jesus Christ, redeemed from among the dead.

Looking back, I recall now that I had absolutely nothing to do with this amazing thing. I didn't go looking for Jesus. I never once thought for a second I would find Jesus or that I even needed Jesus. I went looking for religion and found a Savior. I went looking for ritual and found a Redeemer.

The Lord Jesus saved my life and I will forever proclaim the glories and riches of His majesty.

No Caveats

As several intentions drove this work, my primary task necessarily becomes impressing upon you the fundamental truths concerning the eternal condition of your soul. I seek to invoke examination. Would you not examine the condition of your own heart, your standing before the Lord?

Disparate erroneous conclusions have cluttered the minds and hearts of many warriors. I have sought to dispel these erroneous conclusions, particularly those concerning salvation and one's standing before the Lord as it pertains to the conduct of war. I've been privy to many of these errors and I assure you that they are as dangerous as any historical heresy, turning men from the reality of the risen Savior. Thus, God issues no caveats in either direction.

I pray that you've seen that your occupation has no more to do with your salvation than the color of your skin. The occupation of soldiering merits special consideration in several areas, just not with respect to salvation. Jesus possesses a special love for the soldier. The soldier suffers uniquely as he is confronted with the most difficult of tasks, the destruction of the Image of God that is his fellow man. A soldier bears scars unfamiliar to those who have not borne that same burden.

As God is no respecter of persons, this includes the soldier. Just because the soldier suffers uniquely, God has not established special provision for him to enter into His graces.

I pray that the erroneous and heretical notion that military service uniquely qualifies one for admission to heaven has been soundly refuted. Nothing could be more unbiblical and dangerous with consequences that literally span eternity. I urge you to serve with honor, serve with pride, but serve with the knowledge that you ought to serve something yet higher than even the defense of the nation and reconcile. Reconcile this with your military service and see it for what it is.

I pray that you'll reconcile, that you'll come to terms with the true nature of military service as it pertains to your positioning before a holy God. Understand that reconciliation is not only possible but practically demanded once you have obtained moral clarity and a true understanding of the nature of things, particularly the nature of things in the eyes of God.

God calls all men to stand firm against evil and injustice. Heed the words of Paul as he writes,

> *Put on the whole armor of God, that you may be able to stand against the schemes of the devil.*
> *Ephesians 6:11*

Understand the nature of warfare and evil. Come to terms with the fact that there is evil in the world and that evil will never be appeased and that God ordained godly men to fight, to meet the enemies of Him on the fields of battle, to wage war on behalf of the oppressed, the weak, the vulnerable. Consider the call that God placed upon your life to exactly this thing, the thing that is war. Though you may have joined out of other motives—job training, education benefits, etc.—know that you serve a higher call whether you intended that or not.

Finally, I pray that you have seen that God has little regard for feckless men who stand for nothing, even if it costs them very little. I pray that you've seen that the warrior is a man of honor in the eyes of the Lord, a man who walks worthy of that to which God has called him. Boldness, audacity, courage and strength—as much as these attributes undergird martial endeavors, they ought to undergird the faithful life of the warrior, in all things that we might pursue.

Reconcile these things in your mind and heart and above all, seek after Him.

Rounds Complete

As I contemplate the end of my military service, my wife and I exist in an unusual condition. As much as I am ready to move on with life, as weary as I am of being away from my family, as hard as it has gotten to leave, she desires that I continue to serve. This is not the normal situation. Usually, the wife is urging the man to get out while the man wants to stay in but for us, it is the opposite. She fears that I will not adjust well to civilian life, and perhaps she is right. Words fail to capture my thoughts concerning these last decades of service and war.

How can I ever explain to my wife the love that resonates between men who would, whether of Christ or not, lay down their lives for their friend? How can I ever explain to my wife about the desperation and frustration of the fog of battle, where the easiest things become excruciatingly difficult? How can I ever explain about days like Mother's Day 2006 or the impossibly confusing night that we killed AMZ? I am determined to try.

Departure, though I anticipate it greatly, will be tough. I am unsure of how well I will walk away from this high call, the brotherhood of arms. I am quite certain that any civilian endeavor will pale in comparison to the immense satisfaction I've had at being a small part of a great team of men, of patriots who have raised their hands and said, "Here am I, send me," men who never ask how much, who only ask "When?" and "Where?", men who never count the cost.

The only thing that could ever supplant this high call is the highest of calls, that of the Lord Jesus. God has called each of us to serve, to be a part of the real battle, the spiritual battle that rattles the very foundation of this existence. God has called each of us to a unique path and He has equipped us to walk that path, that we might not stumble. What is God calling you to today? Maybe he is calling you to continue to serve or maybe He is calling you to a greater conflict. Maybe He has a mission for you that you never could have imagined. Would you submit to Him today?

My prayer is that you've seen the utility of war, the necessity of war, the honor and love that undergirds war practiced rightly, that God *teaches* and *tests* through war. And I pray you recognize that war, though brutal, serves the purposes and nature of a holy and righteous God.

Brave Rifle, meditate on these truths. Allow them to speak to your heart. Reconcile them with your own service and come to terms with the very nature of armed conflict as it pertains to your standing before the Lord.

As my time draws near, I've wondered how I will ever explain to Ami what I feel in my heart regarding combat. How will I ever explain to my children that it is with absolute moral clarity and the fullest conviction of the Holy Spirit that I pursue the utter destruction of our enemies—relentlessly, zealously, unequivocally, and irrevocably. To God be the glory, forever and ever.

Amen.

Section 5:
War and the Nations

Chapter 23

June 11th (Iraq)

We finally got busy and executed a couple of targets but the ancillary aspects of this current conflict challenge reason. I resolutely stand by the 'just war' nature of these conclusions. Evil does exist, a great evil that I gaze upon once more, and I stand by the need for strong and courageous men to confront this evil. Complexity threatens to cloud the issue.

Make no mistake, we face a great evil in ISIS. The only difference between this evil and what I've seen in the past is coalescence. This evil has organized and as such, holds different, sometimes competing objectives with like-minded groups. ISIS declared the establishment of the Caliphate, anointed Abu Bakr al-Baghdadi the Caliph, and demanded allegiance from the *Ummah*. Many Muslims feel it is too soon. They aren't ready. Despite lacking consensus, ISIS brazenly presses forward and the more we engage, the more I understand that this is a nation-state. They are currently on the defensive, losing ground on several fronts, but this is a nation-state with organization, governance, and most of all, territory.

Their territory lends legitimacy but at the same time renders them vulnerable. Al-Qaeda and its associated offshoots find power in that they have no center-of-gravity. Over the last 15 years American forces decimated the leadership hierarchy of the movement all the way to the top, and still they function and expand in capability. AQ epitomizes the power of a leaderless organization. United by a shared ideology, a global network of fighters all pursue the same objective, confrontation with the west. Al-Nusra front, the AQ branch in Syria, is in many ways more dangerous than ISIS. They recognize ISIS for their strength but also for their weakness.

It is ISIS's claim to legitimacy as a Caliphate that renders it vulnerable. Fighting AQ is hard precisely because they possess no center of gravity. You find yourself playing a global game of 'whack-a-mole'. It is difficult to develop a cohesive global strategy to combat an ideology. ISIS, on the other hand, is centralized, well-defined, and able to be engaged. There is a well-defined front, defensive positions that separate the good guys from the bad. They wear uniforms, sport flags, and have all the markings of a legitimate army.

It is the strange conglomeration of previously hostile bedfellows—Iran and the Quds forces, Iraqi regulars, Shiite militias, Russians, different Kurdish groups, Syrian regime forces, American military, 'moderate' rebels—that generates angst and confusion. Competing objectives, internal and historical animosity, and temporary alliances of convenience render the outcome decidedly uncertain.

Sin complicates matters on a national level. On the surface, nations declare noble objectives, but every nation possesses ulterior motives. Russia seeks regional influence and prestige, ultimately seeking to bolster their global standing as a superpower. Are they really interested in destroying ISIS for the sake of goodness? One could make a similar claim about American intentions though we at least keep up the pretenses of righteousness in liberating the Iraqi people and allowing them to govern themselves in freedom.

What happens when liberated people choose that which is oppositional to the liberators? What happens when they choose ethnic cleansing or Sharia law, concepts contrary to our ideals of righteousness? We free them to choose and they choose wickedness.

Many denounce warfare for this very reason: that nations seek that which benefits them thus, in their mind, eliminating any potential for righteousness. Does self-interest or self-preservation betray goodness? Are they mutually exclusive?

An analogy might help us reconcile the issue. On a personal level, if I witnessed a gang of thugs robbing a man, I would engage and render assistance. Perhaps this man would offer me a token of appreciation for assistance. Maybe he owns a restaurant and offers me a discount. He tells his friends about me and I now have prestige around the neighborhood, influence. People come and ask me about helping them defend themselves against these same thugs. My defense of the man is a good thing and the ancillary "benefits" to me do not negate the initial goodness.

Let's examine it from another perspective. A gang of thugs robs the same man and I hear about it. Following this, I purchase a gun, install new locks on my home, and buy a large dog. I go to the man and other neighbors and we come to an agreement to confront the thugs and protect one another from them. This benefits me and my family. The man seeks to protect his restaurant so he can feed his own family. We have different intent but could any argue that our measures are not good even though they serve disparate objectives under the guise of unity.

Hatred from sin complicates what is previously clear. To carry it a step further, the restaurant owner comes from Peru and absolutely hates Argentinians. His father died in the war many years ago and he grew up in an orphanage as a result. The gang of thugs is comprised of largely Argentinians, so the man decides to seek out the thugs and burn down their hideout leaving them no place to hang out and plan their robberies. In the fire, three children die and the rest now seek to kill the man in revenge whereas before, they only sought to rob him. Again, the overall objective remains, protect the man's restaurant. However, his hatred springing from the sin in his own heart drives him to act unjustly. Who do we back?

Some days I look to the common grace of God and I thank Him that so many disagree over the meaning of Muhammed's speech at Ghadir Khumm. Were it not for the Sunni-Shia split, I can scarcely fathom how different things might be. Perhaps the best strategy is to keep this place embroiled in a state of persistent conflict with as minimal of an investment from the U.S. as possible. Some inside the Beltway favor a strategy of containment. The current fight costs the government about $12 million a day, a veritable bargain in terms of modern war and tangible progress exists. Can we keep ISIS from engaging outside of the region? That remains the most important question in terms of American interests. The future is very uncertain.

What must the believer do in the face of so much complexity and uncertainty? The believer maintains a relationship with the Lord Jesus Christ and he should rely upon the convictions of the Holy Spirit. If the nation truly conducts actions or pursues objectives that violate a man's Christian conscience, then that man should not partake. However, I would offer that one could boil it down to the basics.

In all of this complexity of our current war/not war, the second target we took down a few days ago was a mid-level financier. He funded evil ISIS operations. Exploitation of his cell phone revealed that he was also involved in child pornography and rape. By the actions of our task force, this man received a just retribution. Of this, I remain intensely comfortable.

23. God and the Nations

Consider that from the motion of molecules to the clashes of kingdoms, the breadth of God's sovereignty defies comprehension. My hope is that I've imparted to you but a sense of His authority in all matters and so we'll return to matters of the state, as war is truly a corporate endeavor. We've examined God's providence, his interaction with the individual soldier, but what of the nations? Does God move among the nations? Does God judge the nations? What of the lone warrior, the pawn amongst kings? How does the Kingdom of God relate to worldly kingdoms?

Of God and Kings

As much as the attributes of love and wrath contribute equally to the fullness of God's character, so too do the concepts of *transcendence* and *immanence*.

God is completely *transcendent*. He is wholly other—uncaused and uninfluenced by His creation, separate in every way. He is apart and above as He simultaneously sustains all things by the power of His very will.

He is *immanent*, intimately involved and concerned with the most minute aspects of existence. He is entirely committed to fulfilling His decree and keeping His promises. Thus, He is personal and through the shed blood of Christ, accessible as the author of Hebrews exhorts, "Let us then with confidence draw near to the throne of grace." (Hebrews 4:16) As a believer, He is my heavenly Father and I may approach him boldly, on my knees in humility, but boldly.

However, as much as He is a personal God, He also deals with the nations, shaping global events in accordance with His sovereign decree. The Psalmist declares God's authority over the nations.

> *Nobles shall come from Egypt; Cush shall hasten to stretch out her hands to God.*
> *Psalm 68:31*
>
> *For kingship belongs to the Lord, and he rules over the nations.*
> *Psalm 22:28*
>
> *All the nations you have made shall come and worship before you, O Lord, and shall glorify your name.*
> *Psalm 86:9*

Nebuchadnezzar affirms God's authority over the nations,

> *all the inhabitants of the earth are accounted as nothing, and he does according to his will among the host of heaven and among the inhabitants of the earth.*
> *Daniel 4:35*

God has always dealt on a national level. Thus he raised Assyria against Israel, Babylon against Judah. He freed Israel from Egypt and subsequently deployed them against Canaan. Today we see just that—God raising nation against nation, king against kings. We see God moving vast groups of people around the earth in a global migration.

Many decry and fear the mass migration of Muslims fleeing conflict in the Middle East. Europe, and increasingly the West, have been forced to handle this enormous influx of people, many of whom do not seek to assimilate and hold values and beliefs antithetical to contemporary Western ideals. Unquestionably, dangerous elements seek to exploit this migration and infiltrate the West posing as refugees. A manner of care and caution is prudent.

How can we deny God's hand at work? Consider another perspective, that of God's sovereignty. God has brought those who don't know Christ from a land where Christ is not preached to nations where Christ may be freely proclaimed.

Consider that I do not have to go to Iraq to proclaim Christ to Iraqis. He brought two Iraqi refugee families to my church in Tennessee. I don't have to go Africa to witness to Somalis. I can make the 45-minute drive to Nashville. Now, please don't hear me say we shouldn't go if God calls us, but do hear me say that we cannot turn a blind eye to the amazing work God is already doing in bringing those who don't know Christ to those who do.

God's providence extends to every level of existence as He governs both the significant and the seemingly inconsequential.

Kings by God

Perhaps most amazing is God's providential concurrence in all of this, that the affairs of kingdoms and nations along with the affairs of individual men run perfectly in line with the decree of God. He not only exercises authority over the nations, but the affairs of nations equally support His will just as the affairs of individual men do.

> *The king's heart is a stream of water in the hand of the Lord; he turns it wherever he will.*
> *Proverbs 21:1*

The water does the work and the king directs it. Notice again, here in the proverb, the concurrence. Ultimately, the LORD directs the will of kings and nations much as he directs the will of men in the spirit of concurrence.

Consider that God directed the signing of the Sykes-Picot agreement in 1916 as much as Saddam Hussein's invasion of Kuwait in 1990. By His sovereign hand, Germany opened a second front in the east and Charles Martel defeated the Muslims at Tours in 732, effectively saving Europe from conquest. All things serve His will from the actions of individual men to the affairs of states, no matter how grandiose or mundane they may be.

Judgment of Kings and Nations

Consider the judgment of God upon the nations. God deployed Assyria against the sins of Israel. Speaking to this, Isaiah writes,

> *Ah, Assyria, the rod of my anger;*
> *the staff in their hands is my fury!*
> *Against a godless nation I send him,*
> *and against the people of my wrath*
> *I command him,*
> *Isaiah 10:5,6*

God raised up Assyria as the rod of His anger. Assyria's fury against Israel is actually God's fury and He has sent them against the godlessness of Israel.

> *But he (Assyria) does not so intend, and his heart does not so think;*
> *Isaiah 10:7*

Assyria intends destruction and plunder. Assyria does not invade Israel out of godly obedience. They intend evil and destruction to satiate their own desires, serving their own evil wills. Nevertheless, God utilizes even their evil will, directs it as a stream of water, turning it wherever He wills. (Proverbs 21:1)

Just over 100 years later, God raises up another nation, Babylon, "that bitter and hasty nation, who march through the breadth of the earth, to seize dwellings not their own," (Habakkuk 1:6) and deploys them in judgment against Judah. Their intent is to seize for themselves. God likewise directs their will as a stream of water, accomplishing His intent, judgment against Judah.

God first offered mercy to Canaan, suffering His people in bondage for 400 years. He then deploys Israel against them as an instrument of both His divine wrath and His sovereign mercy found in the promises to Israel, of which conquest was as much of a part as was His promise to bless all the nations. The will of God, in perfect confluence with both the will of nations and the will of individual men, executes His decree to perfection, accomplishing all that He has predetermined.

Consider that national judgment is perfectly congruent with the fate of individual people in concurrence with the perfectly pleasing will of a holy God. Fall to your knees and join Paul in declaring, "Oh, the depth of the riches and wisdom and knowledge of God! How unsearchable are his judgments and how inscrutable his ways!" (Romans 11:33)

Pawns and Kings

The course of the Church turned in A.D. 313. That year, Emperor Constantine issued the edict of Milan officially ending state-sanctioned persecution of the Church, a centuries-old policy initiated in the days of Trajan and Domitian. The edict eventually led to the adoption of Christianity as the state religion of Rome and ended, for a millennium, the existence of the Church as it had been established.

Contrary to what modern secular skeptics and even many Christians believe, Jesus was the first religious advocate for a distinct separation of church and state. At one point, the Pharisees approach Jesus and inquire, "Is it lawful to pay taxes to Caesar, or not?" (Matthew 22:17) to which Jesus responds, "Therefore render to Caesar the things that are Caesar's, and to God the things that are God's." (Matthew 22:21) Allow me to paraphrase. Jesus says, of course pay your taxes. You are citizens of Rome, so pay your taxes. Don't confuse worldly affairs with the affairs of God. In other words, pay your taxes, obey the laws, and submit to the government.

Peter speaks just as clearly,

Be subject for the Lord's sake to every human institution, whether it be to the emperor as supreme, or to governors as sent by him to punish those who do evil and to praise those who do good.
1 Peter 2:13-14

The Bible calls the believer to submit to the government and in reality, every human institution.

Sacralism is the conjoiner between religion and government. Prior to Christ, every religion was sacralistic in that it existed inseparable from government, a bastardization of religious and state-sanctioned authority. Religion was indistinguishable from the state, lashed up at every level. Islam is a perfect example of a sacralistic religion. Islam governs every aspect of Muslim life, including governance.

Jesus shattered sacralism, completely and definitively delineating spheres of influence and calling believers to be subject to whatever civil authority they might find themselves underneath. Interestingly, as they existed underneath a hostile Roman empire with an official policy of persecution, the church exploded across the known world, unstoppable as it was of God.

However, once the Roman state co-opted Christianity, we immediately saw the introduction of man-made institutions into the church, something never intended. Thus, we see the advent of popes and an unbiblical religious hierarchy. Associated heretical policies led to unbiblical practices such as infant baptism as a form of census taking for taxation purposes, the buying of indulgences to fund church building projects, and the cursed doctrine of papal infallibility. These injustices enslaved countless millions and continue to enslave to this day. Arguably, nothing has done more to damage the Church than fusing it with government.

Interestingly, we observe a familiar pattern in modern day China. The Church, the underground church, explodes throughout China and is growing beyond bounds. By some estimates, China will soon become the largest Christian nation. The official Chinese policy toggles between moderate toleration and outright hostility. In the last couple of years, the government established a series of state-sanctioned churches which sounds like a good thing. Yet, the existence of state-sanctioned churches allows the government to watch and control as they censor messages and monitor what is said and done. This coerced sacralism threatens the foundations of the Chinese church more than any persecution.

The true church is one that is free and independent of government. Like all things, governments exist under the authority of God. Paul writes,

Let every person be subject to the governing authorities. For there is no authority except from God, and those that exist have been instituted by God. Therefore whoever resists the authorities resists what God has appointed, and those who resist will incur judgment.
Romans 13:1-2

We could spend no small amount of time expositing the truth contained herein, but for starters Paul calls upon the believer to submit to the governing authorities as they, the government, get their authority from God. Government exists as God willed it to exist, and to resist this is to resist God.

God instituted government for a very specific reason,

> *for he (government) is God's servant for your good. But if you do wrong, be afraid, for he does not bear the sword in vain. For he is the servant of God, an avenger who carries out God's wrath on the wrongdoer.*
> *Romans 13:4*

God instituted godly governments as his sword and shield, as his servant for good, to visit wrath upon the evildoer. Peter affirmed that governments are "sent by him to punish those who do evil and to praise those who do good." (1 Peter 2:14)

I was a patriot long before I was a believer. I still occasionally battle my tendency to view things in terms of red, white, and blue. I'll make the necessary caveat that I love this country and truly believe it to be a great country yet many today ascribe to America that which was never intended, the favored status of God. Many today believe in their hearts that America's fate is somehow closely tied to the fate of Christianity, and nothing could be further from the truth.

The fact is, in eternity, I'll not profess allegiance to any flag but only to the Lord, Jesus. Some may take issue with this, but as much as I love my country, America will one day be a footnote in the salvation history of man, forgotten and discarded unto the Lord.

The Kingdom of God

We must work to maintain a *distinction* between the Kingdom of God and worldly kingdoms. The tendency to allow nationalism to color just war conclusions is real and dangerous lest we mistakenly bestow upon any nation a favored status.

Lest you be confused, God is no respecter of persons. Paul writes,

> *The gospel is the power of God for salvation to <u>everyone</u>* (emphasis mine) *who believes, to the Jew first and also to the Greek...*
> *Romans 1:16*

Later, He writes,

For not all who are descended from Israel belong to Israel.
Romans 9:6

Christ inaugurated the Kingdom of God, a spiritual kingdom. All those who believe, no matter their ethnicity, gender, nationality, or race are grafted in to spiritual Israel. They are adopted as sons.

There is neither Jew nor Greek, there is neither slave nor free, there is no male or female, for you are all one in Christ Jesus.
Galatians 3:28

The Kingdom of God transcends divisions, both man-made and those divisions bestowed by our Creator. Daniel affirms the primacy of God's kingdom with respect to the nations,

And in the days of those kings the God of heaven will set up a kingdom that shall never be destroyed, nor shall the kingdom be left to another people. It shall break in pieces all these kingdoms and bring them to an end, and it shall stand forever...
Daniel 2:44

As I render to Caesar the things of Caesar and to God the things of God, I embrace the notion that the Kingdom of God is so much more than America or any other nation. The Kingdom of God spans continents and countries. The Kingdom of God destroys borders and boundaries. It supersedes the grandest endeavors of men.

His realm is one of the heart.

Jesus jealously distinguished the building of His kingdom from that of the kingdom of men. Perspective must temper our conclusions. In the *personal* considerations of the ramifications of just war doctrine, we see an end-state in the revelation of the Messiah. This revelation relegates matters of justice and evil as ancillary though important. To the individual, Messianic revelation reverberates in eternity.

Corporately, we must apply the same perspective. As kingdoms rise and nations fall, God is primarily about the business of building His kingdom. All other national endeavors, martial and otherwise, yield to this objective.

This prime cause portends why we see Christ *not* exert certain forms of justice, for the sake of introducing God's preferred realm that much more clearly. Justice must be exercised and evil held to account, but for the sake of his more jealously-devoted kingdom spanning hearts and minds, He Himself occupied the position of carpenter over that of patriot or revolutionary.[87]

As agents of His temporal justice, the warrior must take special care in not confusing the two. Ultimately, a firm grasp allows us to shape our response to the corporate. How must the believer behave?

Just Following Orders

In light of God's call to submit to the governing authorities, inevitably the question arises, "What if my government is evil?"

What if my government has evil goals and intentions?

What if my government is oppressive?

What if my government is not godly?

Am I still called to submit?

The Christian resides and operates in two distinct spheres. I exist as a resident alien, not of this world, and as such, I seek God's will and purpose for my individual life. Accordingly, I pursue God and holiness, love and serve my family, and seek whatever God has in store for me. I ultimately seek out those He has placed in my life that I might speak the words of life to them.

As a citizen of this country, I obey the laws according to my conscience and the word of God. We've determined that legitimate armed conflict is congruent with Christian life. Historically this nation made discernment easier, though that condition diminishes over time.

What I mean by that is, imagine you were a citizen of Germany prior to World War Two? Dietrich Bonhoeffer was faced with just such a notion. That he was eventually executed for plotting to assassinate the Fuhrer speaks to his position before the government. As a devout Christian, how can we reconcile Bonhoeffer's actions with God's call to submit?

The answer resides within the Christian conscience. All believers, if they are truly of Christ, have access to and a relationship with Jesus. The Holy Spirit indwells our very bodies, and He teaches and illuminates, guides and convicts. It is left to the individual warrior to examine and call upon his conscience, choosing to either fight or flee, condemn or submit.

I'd suggest though, that God places the responsibility to resist evil upon the Christian. I'd likewise submit that it is the soldier's responsibility to

---

[87]Kevin Schmidt, *Pacific House of Missions*, personal correspondence.

assess and to resist immoral and ungodly orders. Accordingly, the warrior's duty, as he is a warrior for righteousness *before* he is a warrior of the state, is to without hesitation disobey evil orders despite any consequences.

Adolf Eichmann, the architect of the Holocaust, claimed in a letter requesting clemency that he had just been following orders, that he and others were, "forced to serve as mere instruments."[88] Numerous other Nazi's at the Nuremberg War Crimes Tribunal claimed a similar defense so vigorously that it became known as *Befehl ist Befehl,* literally 'an order is an order', heretofore known as the Nuremberg defense. The Tribunal vigorously rejected this argument though war criminals throughout history, from Abu Ghraib to Bataan, continue to seek shelter in this defense.

My obligation is to the Lord before it is to the state. Were any conflict to arise between my Christian convictions and what the state asks of me, I would have no hesitation in resisting, regardless of the consequences that may occur because of this resistance. I'd suggest though that we ought to resist throwing out the baby with the proverbial bathwater in this regard.

Hugh Thompson provides a perfect example of this point. LT Calley in committing atrocities at My Lai had been, of course, just following the orders of his superior officer CPT Medina, when Warrant Officer Thompson confronted him. As a result of incidents like My Lai and other singular accounts of atrocity, many subsequently condemned the entire war effort in Vietnam. The injustice of atrocity necessitated a condemnation writ large in their estimation.

Thompson, who personally confronted atrocity with all the courage of a true warrior, never flagged in his support of the war. The United States fought in Vietnam imperfectly, but to confront the evil and godless spread of communism. This was a noble endeavor, a godly endeavor, a just war. Individual instances of atrocity and injustice do not necessitate condemnation of the broader conflict any more than the tainting of motivations by sin at varying levels of implementation necessitates condemnation. Thompson acted just as the warrior ought—confront evil in person, yet assess the broader context of the conflict for its greater good.

Bonhoeffer, on the other hand, looked at his government in the broader context and noted the inherent evil of Nazism, and so he resisted in

---

[88] Joshua Barajas, "How the Nazi's defense of 'just following orders' plays out in the mind," *PBS NewsHour,* accessed May 26, 2016, http://www.pbs.org/newshour/rundown/how-the-nazis-defense-of-just-following-orders-plays-out-in-the-mind/.

accordance with the dictates of his conscience and his relationship with the Lord, Jesus.

We must do no less than either Thompson or Bonhoeffer in this regard.

War Without End (A Case Study)

In A.D. 620 at a cave called Hira in the Meccan countryside, the angel Gabriel appeared to a solitary 40-year-old merchant and gave to him the beginning of a revelation, "Recite in the name of the Lord Who created." (Surah 96:1) The merchant went on to proclaim the revelation, urging the local Arabs to abandon polytheism and worship the monotheistic god, Allah. Gabriel appeared repeatedly to the merchant over the course of the next 22 years providing continued revelation which his followers recorded. His name was Muhammed; the writings became what is known as the Quran.

Today, early in the 21st century, a war rages across broad swaths of the globe, a war of attrition, a global war between superorganisms engaged in true Darwinian survival of the fittest. Islam wars with all comers, people of the Book (Christians and Jews), apostates (those who have abandoned Islam), Hindus, Buddhists, Atheists, essentially all infidels (non-believers). This war traces its roots to the 7th century Arabian Peninsula.

Interestingly, Muslims see themselves as of the same religious tradition as Jews and Christians. They trace their lineage to Abraham and revere prophets such as Noah, Moses, and even Jesus. They condemn all aspects of Trinitarianism as polytheism and believe that Christianity perverted the true faith. According to them, Islam is the final correction to the faith and Muhammed the final true prophet.

In the latter years of his life, after the *Hijra* to Medina from Mecca, Muhammed united numerous Arab tribes, sieged Mecca, and began several wars of conquest with neighboring tribes. By his death in 632, he claimed thousands of converts to Islam.

Islam spread rapidly following Muhammed's death, despite a conflict over succession, primarily by the sword. Islam itself means "submission" or "surrender," and per the Quran the faith must be spread over the entirety of the globe by whatever means possible. Non-believers possess three distinct options when confronted by Islam: submit and convert to Islam, retain your faith but pay a *Jizya* (tax) and exist in a state of subjugation to the *Ummah*, or death.

By 750, Islam reached its zenith economically, geographically, culturally, and politically. Muslims conquered the entire Middle East, most of northern Africa, western Asia and were expanding into Europe from the Iberian Peninsula in the West as well as from the East. For nearly 500 years this "Golden Age of Islam" pervaded, and it wasn't until the Mongol invasions of the 13th century that the Caliphate split and Islam began a slow decline. The Caliphate gave way to the Ottoman Empire which dominated for the next several centuries before declining into the "Sick Man of Europe" with its inevitable dissolution following World War One.

Today, Islam claims nearly 2 billion adherents, roughly 23% of the world's population. It is the second largest religion, following Christianity, and the world's fastest growing religion. As of this writing, Abu Bakr al-Baghdadi proclaimed himself the Caliph and declared ISIS as the re-establishment of the Caliphate. Though most Muslims do not acknowledge ISIS and Baghdadi, they march forward undeterred with minds bent toward global domination.

By its very definition, Islam knows no peace. Intellectual dishonesty underscores contemporary pronouncements of Islam as a "religion of peace". Many have called for an Islamic Reformation, for Muslims to abandon archaic practices and assimilate. This appeal to Islamic liberalism is likewise rooted in ignorance as an Islamic Reformation has already occurred.

Religious reformation implies a return to the purity of the particular religion. The Protestant Reformation saw an abandonment of heretical Christian sacralism that perverted the faith and a return to the Pauline theology of the Bible. The Islamic Reformation happened toward the end of the 19th century and in the early part of the 20th. Jamal-al-Din Afghani and his acolyte Muhammed Abduh are largely acknowledged as the founders of the modern Islamic movement. They traveled the Muslim world preaching Islamic unity along with confrontation of and resistance to imperialistic Western powers.

The Muslim Brotherhood, the first and still largest modern Islamist movement, was founded in 1928 and drew upon the beliefs of Afghani and Abduh. The Islamic Reformation has seen a widespread increase in conservative Islamic belief, greater promotion and acceptance of Shariah law, and an accompanying emphasis upon confrontation with the west or in short, a return to fundamental Islam. Sayyid Qutb, an Egyptian Islamist executed in 1966, is largely seen as the modern steward of fundamental Islam or Islamism—with clarity we could dispense with the euphemisms and just refer to it as Islam.

Christians look to Wycliffe, Luther, and Calvin as the stewards of a return to the fundamentals of the religion while Muslims have Qutb and the Muslim Brotherhood in that same regard. "Islam is the solution," and "God is our objective. The Prophet is our leader. The Qur'an is our law. Jihad is our way. Dying in the way of God is our highest hope. God is greater!" These are the slogans of the Muslim Brotherhood and in many ways, the Islamic reformation.

Islam is a religion of conflict, a religion of the sword. This is the uncomfortable truth that many refuse to accept. The sooner we have the moral clarity to understand that the more faithful one is to Islam, the more likely he is to desire to cut off my head, the more appropriately we may respond.

Ishmael and Islam

Peace is a dream, a fantasy when speaking of Islam. As much as an understanding of the religion itself along with a cursory look at history affirms this, so too does the Bible. Arabs, like Jews, find their roots in the Patriarchs.

At one point, Abraham lost faith in the promise that he would bear children through his wife Sarah, and so she gave to him their slave Hagar to conceive a child. It is easy in retrospect to realize that this likely would not turn out well for Abraham on many different fronts as Abraham takes Hagar and 'knows' her and she does in fact conceive.

Hagar began to express open hostility and contempt for Sarah and so she begged Abraham to expel Hagar, which he did. Only God's supernatural intervention in the wilderness saved her life. Appearing to Hagar, the angel of the LORD says,

> *Behold, you are pregnant*
> *and shall bear a son.*
> *You shall call his name Ishmael,*
> *because the Lord has listened to your affliction.*
> *He shall be a wild donkey of a man,*
> *his hand against everyone*
> *and everyone's hand against him,*
> *and he shall dwell over against all his kinsmen.*
> *Genesis 16:11-12*

Hagar returns to Abraham and bears him the son Ishmael through whom Arabs trace their own lineage to Abraham. Later, Scripture records that

Sarah finally bears Isaac, in accordance with the promise of God.

Examine the angel of the LORD's account concerning Ishmael. He was a 'wild donkey' of a man and his hand was against everyone. Everyone fought against him, and he dwelt in a constant state of conflict with his kinsmen, his neighbor. Has anything more accurately depicted Muslims than this proclamation?

Ishmael's descendants are Arabs, a race, not Muslims, a religion. This is true but Islam is a decidedly Arabic religion and almost all Arabs profess Islam. The two have become virtually inseparable though the religion has spread to many non-Arabic parts of the globe.

Paul in Galatians 4 speaks allegorically to the two women and their sons. Hagar the slave bears Ishmael, the child of the flesh and all subsequent children of the flesh. Sarah bears Isaac, the child of the promise, the child of the Spirit. These two children represent the way of the law which is death (Ishmael) and the way of the Spirit which is life (Isaac). Of course we know that the fulfillment of the covenant of grace destroys blood lineage in the line of the children of promise as all who follow Christ, no matter their blood, become children of promise. (Romans 9) Also, as much as the early Pharisees were guilty of legalism, seeking to impose righteousness through rabid enforcement of the law, Islam is an entirely legalistic system, completely based upon rigid enforcement of strict laws to attain righteousness.

Islam confronts, violently if needed. It's what Islam does by its very nature.

At the national level, Islam presents the west with decidedly unpalatable options. Do we support the pro-western but brutally oppressive dictator or the freedom-loving (sharia-loving) rebels who will eventually want to cut off our heads? The current civil war in Syria illustrates this conundrum precisely.

Assad is a historically pro-western leader but he is also an oppressive and brutal tyrant toward his people, suppressing human rights and crushing dissent. The original rebels are what we now know as ISIS and Al-Nusra front, a branch of Al Qaeda, both deeply hostile to all things western. Initial American attempts to find and support moderate (read: secular) rebels failed greatly though a new coalition largely composed of Kurds with some moderate Arab support is finally turning the tide as the Kurds are largely pro-western secular Muslims.

We must realize the spiritual foundations of this battle. I truly believe that an angel appeared to Muhammed but it was definitely not the angel

Gabriel. Can you think of a more effective system for enslaving billions of people, keeping them in bondage apart from the Gospel of Jesus Christ? Islam is as satanic a notion as any, and it is with full assurance that we must realize that it was a fallen angel, perhaps Satan himself that delivered this evil system to Muhammed and the world.

We must realize that three things stand between your family and the sword. First, the spread of Islam relies upon the piety of faithful men, something that apart from the Holy Spirit the Bible assures us just doesn't exist. Men's hearts are fickle and very few men will subordinate themselves to a cause that will definitely run in opposition to their own welfare at some point. Apart from demonic possession, it's tough to convince that many men to sacrifice themselves on an altar other than that of their own desires, despite promises of eternal gratification. Men are just not that pious.

Second, disputes over leadership following the death of Muhammed led to the Sunni-Shia split. Today, the vast majority of Islamic violence is Muslim on Muslim as Sunni's in particular consider the Shia as apostate, worse than an infidel, though the Shia also commit great atrocities against the Sunni. This rift divides the heart of Islam and as they war amongst themselves, provides a level of protection to the infidel.

The third thing that stands between your family and the sword is your own sword and shield, the heart of the warrior. I pray you'll not grow weary. This enemy cannot be appeased.

24. Problem of a Godless Army

The nexus of this endeavor to collude theology and military theory resides in a look at the institution I know best, the United States Army. I love the Army. As I stated previously, I've always felt called to be a soldier and it has been an absolute joy to serve this nation, though in confession, as any soldier will attest, things were not always so delightful.

When my 1SG had us report at 0430 for a physical fitness test, I confess that I was not the most joyous of persons. When my battalion lost a sensitive item and we were on lockdown for nearly a week and I received at least a daily beat-down from a two star general, I was not the most joyous of persons. (If I have PTSD, it's from this incident!) When the temperature pushed well past 100 degrees for the third consecutive day during my most recent field training exercise, I was not so joyful, but in general, it has been my privilege and honor to serve. I cannot think of a more noble profession, the profession of arms.

So we will take a look at the nation's oldest and largest institution, and apply a critical eye. As much as I love this Army, very serious issues permeate the ranks with potential grave consequences toward the future of this nation. As relevant as they are for this day, I believe that we may likewise glean some timeless truths from this assessment, truths valid for all time concerning the employment of an army.

The Soul of the Army

Does an institution have a soul? Does it have a heart? The United States Army resonates with a pulse and a life of its own, born in June 1775 of the crucible of conflict as the Continental Army. From Valley Forge to Vietnam, from Antietam to the A Shau Valley, the Army has faithfully upheld its motto, "This we'll defend."

As the land-based branch of the U.S. Armed Forces, Section 3062 Title 10 US Code defines the purpose of the Army as,

- Preserving the peace and security and providing for the defense of the United States, the Commonwealths and possessions and any areas occupied by the United States,
- Supporting the national policies,
- Implementing the national objectives,
- Overcoming any nations responsible for aggressive acts that imperil the peace and security of the United States[89]

The United States Army is one of the largest and most diverse organizations in the world. From a peak of over several million active duty soldiers during World War Two, the Army in 2016 numbers just under 500,000.

Citizens of every state and territory serve. Pick a race or nationality, and you'll find them represented in the Army. I've personally served alongside soldiers from Zimbabwe, Guyana, Poland, England, Egypt, South Africa, Nigeria, South Korea, and Brazil, just to name the ones I can remember. Often, military service provides an easier path to citizenship.

The Army spearheaded social change, desegregated by President Truman in 1948, and led the way for full integration during the civil rights campaign of the 1960's. Today it is one of the most ethnically diverse organizations in the world, and since the 1980's it has been manned exclusively by volunteers, an amazing consideration.

The Army operates as a hierarchical meritocracy. It maintains a strict and rigidly enforced chain-of-command whereby promotion is based solely upon manner of performance and potential to serve at the next higher level. Periodically, Army leaders issue sometimes controversial injects into the promotion system intended to assist certain demographics at predetermined times. In general, promotion is based upon merit and potential.

This is what makes the Army such an amazing organization, the fact that you can progress as far as you are able. You can literally, "Be all that you can be." The only limiting factor is you. The Army will not only allow, but will drive you to maximize your potential.

The Army is an ever-evolving institution, necessary due to the ever-

---

[89]Title 10 U.S. Code, accessed May 21, 2016, http://uscode.house.gov/view.xhtml?path=/prelim@title10&edition=prelim.

changing nature of the threats to the nation. Early in the 21st century, much organizational discussion centered upon the notion of a learning organization. Was the Army a learning organization? Is Al Qaeda a learning organization? Multiple publications and articles assessed the merits of a learning organization, its utility, and whether the Army was a learning organization. Students at the Command and General Staff College at Fort Leavenworth and at the U.S. Army War College at Carlisle Barracks debated the notion. Could the Army learn as an organization? What did that even mean?

In 2006, Brafman and Beckstrom published *The Starfish and the Spider: The Unstoppable Power of Leaderless Organizations* which again set off a series of professional forums and discussions assessing the concept of a leaderless organization, a concept that our enemies seemed to employ so effectively. Was there applicability to the Army?

Most recently in 2015, Wong and Gerras of the Strategic Studies Institute published an article entitled *Lying to Ourselves: Dishonesty in the Army Profession* whereby, after some fairly thorough and pointed research, they called into question the integrity of Army leadership. They noted specific repeated instances of integrity breaches so pervasive that they are never questioned. All officers lie about certain things, things generally deemed of lesser importance.

"Captain, did your unit complete its mandatory sexual harassment training?"

"Without a doubt, Sir!"

One of the factors that Wong and Gerras noted was the sheer volume of things asked of junior leaders in the Army. If you assessed the minimum annual additional training requirements of Army Regulation (AR) 350-1 you would quickly realize that this alone required more training days than existed in a year yet units continuously reported, "all complete," noting that, "the Army is quick to pass down requirements to individuals and units regardless of their ability to actually comply with the totality of the requirements."[90] Never mind training their units for combat. Someone must be lying somewhere.

As it is, I don't ever remember a time when the Army has not been introspective, always pursuing efficiency and effectiveness. From my

---

[90] Leonard Wong and Stephen J. Gerras, "Lying to Ourselves: Dishonesty in the Army Profession," *Strategic Studies Institute*, accessed May 21, 2016, http://www.strategicstudiesinstitute.army.mil/pdffiles/PUB1250.pdf.

earliest days at the Military Academy to my nascent days in the Army, introspection and analysis were always the order of the day. Assemble a group of Army leaders and they will most assuredly end up discussing things that are right and things that are wrong, what the Army should be doing, and what the Army should stop doing. Much of the discussion is normally negative and interestingly, I've noted specifically that junior officers talk about strategy while generals talk about squads and tactics.

As Army leaders have always sought efficiency and effectiveness, much professional discussion concerns sources. What are the sources of organizational dysfunction and inefficiency? What are the primary causes of fundamental issues? All leaders seek to peel back the layers and ascertain whether dysfunction is episodic or indicative of a deeper underlying issue and what that issue might be. I can imagine leaders in antiquity such as George Washington or Napoleon doing this very thing.

I have not had the best of seats to survey the Army until recently. My first four years in the Army I served overseas in Korea and Honduras, assignments that shielded me from much of the inner-workings of the divisional Army. Following that, I served many years in the SOF community, an absolute fantasy land. Those years, particularly once the war started, were a blur of training and combat. Now, the SOF community sought change and evolution at a rapid pace, but in a vacuum. Organizational structure shielded SOF from the happenings in the 'Big Army' of the divisions. I knew it existed, the Army that is, but that was about it. Other than some isolated accounts from a few army buddies, I existed oblivious to the vast majority of the Army. That was about to change.

In 2013, the Army sent me to a Regular Army battalion for my first ever divisional assignment and to say that it opened my eyes was a mild euphemism. Though I truly loved my nearly three years in the unit, I never expected to encounter some of the affliction that has unfortunately become so commonplace.

Yet, as I said, I loved it. I loved the soldiers, earnest young men and women, many of whom truly wanted to make a difference in this world. It wasn't lost on me that nearly all of them enlisted after September 11th knowing that a deployment to combat was likely imminent at some point. I loved the NCO's, the sergeants, a magnificent group of seasoned and respected leaders who literally served as the glue that held all things together when they seemed most likely to fall apart. I loved the officers, a group of committed young patriots—bright, energetic, and motivated. The lieutenants and captains are light-years ahead of where I was at a similar point in my career.

Yet, something was wrong. I could tell from the beginning, but could not diagnose the issue. I observed and watched. As a Battalion Commander and then the rear detachment Brigade Commander of over 3,000 soldiers, I had a front-row seat to the action. And as I observed, I deliberated. Again, something was amiss, but I could not quite fathom what. Lots of symptoms, but what was the source?

Walking around, you would never know something was wrong. Soldiers looked just about the same as they always looked. Our Brigade performed magnificently in combat, yet three things consistently captured my attention, the sheer quantity and scope of affliction within the ranks, the yeoman's efforts of commanders and 1SG's in addressing the affliction of the masses, and the rampant thoughtlessness of *some* of the senior leaders.

Clarifying the Issue

The Army has always prided itself on being a values-based organization, insisting upon integrity at all levels. The Army Values of,

- Loyalty
- Duty
- Respect
- Selfless Service
- Honor
- Integrity
- Personal Courage

are drilled into the heads of new soldiers from the very first day of basic training. One look at the Army Values and you can quickly note that these are good things, all necessary to build trust, an essential component in combat and to a self-professed values-based organization. Interestingly, the Army Values could have been lifted straight from the pages of Scripture. Were I to biblically expound the attributes of a soldier, they would probably look a lot like the Army Values.

Further, the Uniform Code of Military Justice (UCMJ) supports the ethos of the Army, reinforcing that which the Army declared important. Thus, things that may not be an issue in the civilian world will quickly have you running afoul of Army leadership. As an example, the Army maintains, for now anyway, a well-defined sexual ethic. Adultery violates the UCMJ.

Now, it's exceedingly difficult to prove adultery as a commander literally needs a confession, a positive blood test on a child, or some other 'proof' of the act. Yet, it is against the UCMJ and soldiers may be punished if found guilty of committing adultery.

Again, this and other aspects of the UCMJ could have been lifted straight from the pages of the Bible. The Army Values, the Warrior Ethos, the UCMJ: these all serve as a rudder, guarding the hearts and actions of soldiers and likewise, serving as the anchor for the entire organization. Against this backdrop, I've labored over company commanders, 1SG's, and the impossible tasks the Army demands of them.

We had a senior Army leader visit the installation, and as a Brigade-level leader I was privileged to attend a forum with him and the other commanders. He remarked that every single commander and 1SG on every single installation that he visited remarked that the Army required way more of them than there was time to do. An issue as long as I've been in the Army, the literal requirements placed upon company level leadership greatly outpaced their capacity (time) to complete these requirements.

Wong and Gerras drew upon some of this analysis as for years, the Army reported 'all complete' on these extraneous requirements out of one side of its mouth yet with the other, complained that there was not enough time in the day to accomplish nearly all of the necessary tasks. Well, which was it?

This particular senior leader believed the issue to be one of prioritization. The commanders should prioritize their requirements, obtain buy-in from their leadership, and then execute what they are able. Allow their senior leaders to provide top-cover on requirements deemed untenable or unnecessary based upon time and resources available.

This may brief well, but as he spoke I wondered to myself if this would work. I never did such a thing as a company commander. As a battalion commander, none of my company commanders had done such a thing. As a rear detachment brigade commander, none of my battalion commanders had come forward in such a fashion. Every unit I've been a part of reported 'all complete' when it was just never feasible to have completed but a fraction of the required tasks.

In assessing this senior leader's remarks, I determined that most company commanders probably don't have time for that level of assessment. Company commanders exist on the front lines of leadership, literally where the rubber meets the road. To take time to make a comparable assessment is just not feasible for most. Perhaps it should have been the higher headquarters that made this type of assessment. Besides, most things seemed like a priority. How would you prioritize when your headquarters deemed numerous competing demands as priorities, especially when they all seem important?

I had a company commander sleeping at the CQ desk for several nights in a row keeping watch over an imminently suicidal soldier. I had a company commander literally spend several sequential days dealing with a love triangle in the motor pool. I had a company commander make so many trips to the local mental health facility where we sent struggling soldiers that he was quite literally known by name.

These are all good things in that commanders get paid to take care of soldiers and concern themselves with their well-being. On the other hand, we also pay commanders to accomplish the mission, to close with and destroy the enemy, to fight and *win* our nation's wars, and every hour, every day, every week spent dealing with the litany of soldier issues is time NOT spent preparing their units for the rigors of combat. Here is the issue.

The Army standard is clearly a godly standard, whether intentional or not, though I believe it to be intentional. The Army, as it reflects the composition of our society writ large, is clearly and increasingly a godless organization. The primary challenge facing leaders in the modern Army of the United States is this,

motivating godly behavior from the godless, apart from God.

Herein lies the crux of the leadership challenge with which the Army wrestles on many different levels. Allow me to explain. The challenge is, *motivating godly behavior* (compliance with UCMJ and internalization of the Army Values and Warrior Ethos) *from the godless* (majority of soldiers), *apart from God* (in an environment hostile to the intentional proliferation of the Gospel).

My previous commander, one of the greatest combat leaders I've ever had the privilege of serving under, remains convinced that the United States is poised for a fall and that we will lose our next war. I am not sure if I concur. However, virtually every issue the leaders of our Army grapple with stems from the impossibility of the above paradigm. One can only coax a certain measure of godliness from the godless. Failure is inevitable in this regard.

Frankly, the issue is sin running in opposition to values and codes which generates all manner of personal affliction in the lives of soldiers. Sin necessitates mandatory training in an attempt to handle the issue from a secular standing. The resonate sin in our force, not under the conviction of the Holy Spirit or increasingly not even under the common grace of the proliferation of the Gospel, consumes our force and its leaders, diverting them from the most important of tasks—preparing our forces to confront the evil of this world.

Chapter 25

25. Danger of a Godless Army

A Sinful Heritage

I'm not so naïve to think that there was ever a time whereby the vast majority of soldiers confessed Christianity. Soldiers have always been a rough and rowdy crowd. My very first 1SG chewed tobacco, cussed like a sailor, and drank beer every single night. However, he was a great 1SG and I remember him fondly. One of my first instructor pilots was a prolific womanizer, a true philanderer. He was always chasing a different woman and he was a great instructor pilot. We logged many flight hours together. My very first battalion commander was an absolute party animal who had an affinity for the juicy girls in the 'Ville, yet he was a highly esteemed commander.

The barber shops in Korea used to be called the *Steam and Cream* for a very specific reason as they offered, overtly, a very specific extra service for a small fee. This happened on base! The Officer's Club at Fort Campbell used to have actual strippers...on base! Rampant drunkenness used to accompany just about every Army social function, often on base! Sin was rampant, overt, tolerated and even celebrated.

None of this behavior would be tolerated in today's Army. Strippers no longer frequent the Officer's Club—the Army has shut the door on all officer's clubs and were an officer to take his subordinates to a strip club, he'd quickly find himself out of work. Prostitution, as it assuredly exists or even prospers, has gone decidedly underground on military bases, no longer officially tolerated. Unit functions are now much more likely to include family-friendly events, games and such, rather than alcohol. In light of the moral rightness of moves such as this and others, how do we reconcile this with any claim of increasing godlessness? Would not the fact that overt sin such as this is no longer tolerated drive us toward the opposing conclusion?

Godless Nation, Godless Army

"The sky would fall," we were told reference the end of the military's "Don't Ask, Don't Tell" (DADT) policy. DADT was introduced in 1994 by the Clinton administration to permit homosexual soldiers to serve as long as they kept their orientation to themselves. This overturned the military's total prohibition against homosexuality and in reality, allowed the Clinton administration to skirt the issue, claiming allegiance with the homosexual community while at the same time preserving ties with those opposed to allowing homosexuals to serve openly, an influential group in 1994 though not so much anymore.

Ten years into the current war, the Obama administration began to openly maneuver to repeal DADT against strident opposition from numerous factions.

"It would affect readiness! The troops will resist! There will be chaos in the ranks!" Some in the old guard would not die quietly.

Nevertheless in 2011, President Obama, Secretary of Defense Panetta, and Chairman Mullen sent confirmation to Congress concerning readiness and the repeal became law. DADT was history. The homosexual community rejoiced alongside their progressive comrades. Conservatives braced themselves for the impact, the fallout…and kept on bracing. The impact never came. Now, nearly six years later, there has not been a single reported incident concerning the repeal. Why the disparity between perception and reality?

I knew, from the moment it was first discussed that, 1) repeal was imminent and 2) it would have minimal effect. Consider who constitutes the majority of the military: 18-24 year old young people who have been raised in the public school system, likely by secular, unchurched or de-churched parents for whom homosexuality is just not a consideration. They view it as acceptable like any other lifestyle. Who you sleep with is your business. It's what they have been taught. Only the stodgy old guard resisted.

The most senior leaders, as slaves to the progressive civilian leadership of that day, kept any objections to themselves. Similarly, mid-grade leaders such as myself kept our mouths shut lest we endanger our livelihood. And besides, upon what grounds or basis would we ever object?

If the vast majority of soldiers accept it, and they do, then on what basis would we object? The military is a basically pragmatic organization, and if this aspect of morality had no impact on mission readiness, then what's the point? Proponents even recrafted the issue in terms of civil rights and further bolstered their position while minimizing dissent. Who would dissent and be labeled a bigot? In my unit, one senior warrant officer, one, submitted his retirement packet in protest though he retracted it just as quickly.

As America is the great Melting Pot, the military has always reflected Americana. The military traditionally imitates American values, composition, and culture though it maintains a decidedly conservative slant. As society goes, so goes our military. And in some ways, the military lead social change, sometimes for good and sometimes, well...

Spiritually speaking, America is but a shadow of its former self. Though the vast majority of Americans still claim to be Christians, our behavior betrays us. Upon closer examination, it is easy to see that America is a post-Christian nation. Barna Group has been researching Christianity in America since 1984. In a recent study, they conducted over 60,000 interviews using 15 metrics to assess the spirituality of America. What they discovered is not surprising. In 2013, 37% of Americans identified as post-Christian based upon the given metrics. By 2015, that number had risen to 44% as America becomes increasingly irreligious.[91]

A 2014 study likewise yielded a number of un-shocking truths:

- The number of unchurched people in America would constitute the 8th largest country in the world (156 million).

- In the past decade, more Americans have become churchless than the total population of Canada and Australia.

- The majority of the unchurched have attended church previously and could be more accurately labeled as de-churched though the number of actual unchurched people, those who have never attended church, is on the rise.

- The majority of the churchless in America claim Christianity as their faith.[92]

To clarify, church attendance does not make a Christian, as I pray we've covered in depth in the previous chapters. The Bible stipulates church attendance for the believer, fellowship with other believers, and it is a good

---

[91] Barna Study, accessed May 22, 2016, https://www.barna.org/barna-update/culture/728-america-more-post-christian-than-two-years-ago#.V0FU6f7_Opo.

[92] Barna Study, accessed May 22, 2016, https://www.barna.org/barna-update/culture/698-10-facts-about-america-s-churchless.

indicator of spirituality and spiritual growth. The Bible knows nothing of a Christian faith lived in isolation from other believers, from the Church, the body of Christ.

A vast majority of Americans claim, "I'm a Christian, I just don't attend church," or "I'm spiritual, I do church on my own." Jesus would have no idea what they are talking about. Being churched is a good indicator of the spiritual health of both a person and a culture.

Today, most of America is de-churched. Maybe they were raised in church or grew up attending church. At some point, they walked away. Growing in number are the true unchurched, those who have never attended. Their parents may have been churched or were de-churched, but for whatever reason they walked away. A newer group that stands to shape America further is the second generation unchurched, those raise by unchurched parents or guardians. The number of second-generation unchurched will inevitably outpace the other groups

Why does it matter? Let's discuss common grace, the common grace of the Gospel of Jesus Christ. Jesus says to the Apostles that

> ...he (God) makes his sun rise on the evil and on the good, and sends rain on the just and on the unjust.
> Matthew 5:45b

God sends his blessings of sunshine and rain on everyone, not just His people. The common grace of God blesses all men in some fashion. Thus the proliferation of the Gospel serves to restrain sin. Wherever Christ is preached, things are just better. The individual responses of the people, in a way, become irrelevant at the corporate level. The proliferation of the Gospel serves man through the common grace of God.

Thus, as the Gospel is suppressed by the de-churching of America, things will progressively trend downward. The restraint on sin will diminish and the wickedness of men will flourish. According to Paul, speaking on the unrighteousness of men,

> Though they know God's righteous decree that those who practice such things (sin) deserve to die, they not only do them but give approval to those who practice them.
> Romans 1:32

Paul says that men will inevitably not only exchange the worship of the Creator for the worship of the created thing, but will also approve and even celebrate those who immerse themselves in their sin. Consider that the President actually called NBA player Jason Collins to congratulate him on coming out as the first openly homosexual player. "I couldn't be prouder," was President Obama's gushing tribute.[93] Michael Sams' announcement as the first openly gay NFL player likewise met with fawning support from celebrities and government officials.[94] ESPN awarded the Arthur Ashe courage award to Bruce Jenner for his 'transition' to being a woman.[95]

Because of this, the widespread proliferation and celebration of sexual sin including homosexuality, many believe that God will judge America. According to Romans 1, God has already judged America.

Paul writes that, "for this reason God gave them up to dishonorable passions." (v. 26) The *reason* God did this is that they abandoned Him even though they knew Him as all of creation testified to His existence. America turned from God some time ago, and as such God has given us over to our dishonorable passions which Paul further defines,

> *For their women exchanged natural relations for those that are contrary to nature; and the men likewise gave up natural relations with women and were consumed with passion for one another, men committing shameless acts with men and receiving in themselves the due penalty for their error.*
> *Romans 1:26-27*

God will not judge America for homosexuality. God already judged America for worshiping the created thing rather than the Creator. The proliferation of homosexuality speaks to this; homosexuality is God's judgment on America. He has given us over to our sin and we will receive "the due penalty for (our) error."

The homosexual lifestyle, even cursory research demonstrates, brims with licentiousness, decadence, and affliction that weakens communities and our nation. Now it is mainstream in our society, celebrated by leaders and celebrities, and promoted by our institutions. Less than a year after the repeal of DADT, the Department of Defense established June as "Pride"

---

[93] Steve Chaggaris, "'I couldn't be prouder' of Jason Collins, Obama says," *CBS News*, accessed May 24, 2016, http://www.cbsnews.com/news/i-couldnt-be-prouder-of-jason-collins-obama-says/.

[94] "Michael Sam earns national praise, support," *St. Louis Post-Dispatch*, accessed May 25 2016, http://news.live.stltoday.com/Article/360104-Michael-Sam-earns-national-praise-support.

[95] Phil Helsel, "Caitlyn Jenner Receives ESPY Arthur Ashe Award for Courage," *NBC News*, accessed May 24, 2016, http://www.nbcnews.com/news/us-news/caitlyn-jenner-receives-espy-award-courage-n392911.

month. Though troubling, this is all merely symptomatic and shouldn't surprise in light of the de-churching of America. Why would a nation straying from God not proliferate sin, sexual and otherwise?

The de-churching of America generates another symptom, particularly as second generation unchurched people grow to adulthood. Biblical literacy declines yearly. Previously, though one may not have been of Christ, they spoke the language, having been raised in church and Sunday school. When you spoke to them of sin and repentance, they knew the language you were speaking. They understood what you were talking about.

Increasingly today, biblical concepts such as these are not in people's vocabulary, not a part of the vernacular.

"Sin? What is that?"

"What is repentance? Repent from what?"

"Doesn't God love me for who I am?"

Sharing the Gospel message has changed. A different paradigm confronts the evangelizing Christian. We must account for the new context, and the younger a person the more likely he or she is illiterate in the basic aspects of Christianity.

And this is our Army. Our Army is a representative organization primarily composed of young 18-24 year-old men and women, the clear majority of whom are at a minimum de-churched, with a growing number being unchurched or even second generation unchurched. This is our Army and the darkness runs deep along with the inevitable affliction.

A Snapshot

On September 1, 2015, the 39th Chief of Staff of the Army, General Mark A. Milley dispatched a message to the Army saying, "We have the most skilled, ethical, and combat hardened Army in our Nation's history."[96] He goes on to lay out his three priorities: readiness, the future force, and taking care of soldiers, all valid and carefully thought out. His very first statement drew my attention.

We possess a highly skilled army, sometimes too reliant upon civilian expertise, but highly educated and well-trained. Years of persistent warfare have definitely hardened our army. However, the percentage of active-duty

---

[96] Accessed May 23, 2016, https://www.army.mil/article/154803/ 39th_Chief_of_Staff_Initial_Message_to_the_Army/.

soldiers with combat experience continually declines. Moreover, one could argue that after a certain time, combat experience (hardening) becomes combat fatigue, a detractor from effectiveness.

His comment that our Army is the most ethical in our Nation's history is what struck me. Is our Army truly more ethical than it has ever been?

As of the publication of General Milley's message, I had 19 soldiers in my brigade under investigation for rape or sexual assault: rape of a friend, rape of a child, rape of their own child, even rape of their own special needs child. We were in the process of breaking up a marijuana ring in one battalion. We'd just had our second suicide in a span of a few months. Both soldiers hung themselves with their belts in their barracks room. Just a few months prior to General Milley's announcement, I had nine domestic violence cases in one month.

Handling these issues consumed us at some point. We even formed two separate committees that met monthly whose sole purpose was to handle the affliction of our soldiers, as we desperately sought to keep them from self-destructing. Soldiers spent so much time at the local mental health facility that the military health care system began cutting them off, something I'd never seen before.

This is the most ethical our Army has ever been? Now, as my scope of purview has increased, perhaps that has colored my conclusions, but I just don't remember any of this from my last foray into the big Army.

To deal with this glut of affliction, the Army lays these burdens on company commanders and 1SG's. Interestingly, General Milley's number one priority was readiness, as it should be. Could we marshal our forces and deploy them to combat? Were they ready? Soldier affliction due to the increased godlessness of the force and the corresponding increase in rampantly sinful behavior works directly against this objective. Commanders and 1SG's find themselves caught in the middle, straddling priorities. At some point, they have to actually train their forces.

Army Regulation 350-1, table G-1, defines Department of the Army, Headquarters Level training requirements for units. Many are obvious: Army Warrior Training, marksmanship, physical readiness. An army should be doing these things. Others are equally as important, but not as obvious: Anti-terrorism training, Operational Security, Law of War, Personnel Recover, Information Assurance. Again, these are all good things, just not as obviously necessary for the warrior.

Still others exist solely as a secular response to the rampant sin in the force: Alcohol Substance Abuse Prevention, Suicide Prevention, Combating

Human Trafficking, Equal Opportunity, Sexual Harassment and Assault Response Prevention (SHARP), Resiliency Training (though not stipulated at DA level, resiliency training is a mandatory monthly event). Consider that we actually hold classes to teach soldiers not to rape people!

Not only must commanders deal with the sin of the force in handling the immense number of personal issues generated by sinful behavior, they must also train the entire force as a response and in an attempt to prevent this same sinful behavior. Somewhere in there, they must find time to accomplish the mission essential training to prepare the unit for combat.

A 2002 War College study determined that all mandatory training would require 297 training days in a year. Regrettably, each year contains only 256 training days.[97] It is no wonder that commanders and 1SG's feel increasingly overwhelmed at what has been asked of them.

Based upon the deluge of requirements placed upon company level leadership, combat-focused training often takes a back seat to administration and dealing with soldiers' personal issues. Maybe my last commander was correct in ascertaining that United States is set for a fall, capable indeed of losing its next war.

---

[97] Wong and Gerras.

Chapter 26

26. Sex in a Godless Army

We simply cannot address the affliction of soldiers without addressing the penis and its proper usage, rather the implications of its improper use. We'll address other afflictions wrought by pervasive godlessness, but sexuality rules the day. God made us as sexual beings. Sexuality permeates our very existence and as such, it must be accounted for.

Only a few dabble in a singular pervasive sin as afflictions imbricate and become indistinguishable one from another. Generally speaking, most of the afflicted suffer from a multitude of assaults upon their being which is to be expected.

A Godly Ethic

Like our nation, our soldiers are entirely confused about sexual ethics and what godliness in this arena looks like. Lines continue to blur and many descend truly into what would have been deemed utter insanity only 20 years ago. God clearly defines gender and sexuality leaving absolutely no room for vacillation.

The Bible records that the Pharisees confronted Jesus regarding divorce. His response addresses a number of pertinent issues.

Have you not read that he who created them from the beginning made them male and female...
Matthew 19:4

Without hesitation, Jesus refers directly to the first decree in Genesis,

So God created man in his own image, in the image of God he created him; male and female he created them.
Genesis 1:27

God made them male and female—separate, different, not just good, but very good. (Genesis 1:31) Jesus goes on to reference Genesis once more saying,

> ...*Therefore a man shall leave his father and his mother and hold fast to his wife, and the two shall become one flesh'? So they are no longer two but one flesh.*
> Matthew 19:5

It is a true statement that Jesus never expressly condemned homosexuality. He also never expressly condemned bestiality, or pedophilia, or incest. He did promote marriage, solely between a man and a woman, as He likewise condemned adultery, lust, and by inference, fornication. Jesus condemns any sexual activity outside of that between a man and a woman in the context of a biblical marriage. (Matthew 5) The rest of Scripture, including the Old Testament and Peter and Paul in the New, condemns sexual immorality, including any sexual activity outside of a biblical marriage between a man and a woman. Only an extremely painful and distinctly dishonest exposition of Scripture will yield any conclusion other than this.

Some points,

1. God made the male and female, separate.

They are different, distinct. This distinction is a good thing. God made them in His image, unique and with different passions and purposes. To the man, He gave the mandate to work, to have dominion over creation in the Garden, and to teach His wife what God had given directly to Him. God creates woman as man's 'helper'. (Genesis 2:20) Lest any think this a derogatory or diminishing term, consider that God frequently refers to Himself as man's 'helper'. (Ex. 18:4, Psalms 33:20, Psalms 70:5) Woman is to come alongside man and 'help' him in exercising dominion over creation.

God designates separate roles for the man and the woman. First, the man is to serve as the spiritual leader, to have authority over his wife and family. Immediately we begin to encroach upon contemporary thought patterns. The Bible speaks clearly on the matter. Nowhere does God direct the man to rule or dominate the woman, this is solely of the Curse. (Genesis 3:16) God does direct male leadership.

Many pull Ephesians 5:21 out of context when Paul speaks about believers "submitting to one another." They propose a sort of mutual submission between the man and woman. However, this is a weak assertion tempered by fear of rejection and poor exegesis, clearly not the godly intent.

In verse 21, Paul does tell Christians to submit to one another, but then he goes on to clarify what he means. "Wives, submit to your own husbands, as to the Lord." (v.22) "Children, obey your parents in the Lord, for this is right." (v.6:1) "Bondservants, obey your earthly masters." (v.6:5a) Paul says, everyone submit to one another and here is how:

1. Wives submit to husbands,

2. Children submit to parents, and

3. Slaves submit to masters.

Long ago, western society applied a derogatory connotation with the concept of submission. The Bible never quibbles. Nowhere does God call the husband to submit to his wife. The husband is the head of the household, indeed the head of his wife, "so also wives should submit in everything to their husbands." (v.24) Peter exhorts, "wives, be subject to your own husbands." (1 Pet. 3:1)

This straightforward biblical concept generates intense and immediate ire. Antiquated, masochistic, chauvinistic—the notion of male headship drives those who object into a veritable frenzy. They simply cannot believe that someone would advocate such a blatantly discriminatory and hateful belief. I understand how some might object without considering the full counsel of God's word on the matter.

Consider the call given to the husband. "Husbands, love your wives, as Christ loved the church." (Ephesians 5:25) A husband is to love and honor his wife in the same way that Christ loves the church—to die for her, literally. When taken in this context, the notion of mutual submission seems like a paltry concession for the sake of bristled sensibilities. Consider the power of husbands loving their wives as Christ loved the church and wives submitting to husbands in a godly fashion.

Again, submission does not imply subjugation or lack of equality or worth. God does call all Christians to submit in some way. God calls all believers to submit to the government (Romans 13:1, 1 Peter 2) and to the leadership of the local church (Hebrews 13:17, 1 Peter 5). It is the wife that God calls to submit to the husband "as to the Lord", not as if he *is* the Lord. Her ideas, thoughts, and beliefs possess merit; she is just under his headship.

God said that this is the best way. The curse from Genesis 3 opposed this godly pattern from the very beginning. Resistance or outright refusal is merely symptomatic of this curse. God made them male and female, in His image, with different roles, but of equal value. This is the essence of biblical complementarianism, a foreign concept to the unchurched and increasingly, even to the churched, as popular unbiblical thought patterns continue to invade the body of Christ.

2. **Marriage has the mission of procreation and discipleship.**

The very first command given to the *couple* was "Be fruitful and multiply and fill the earth and subdue it." (Genesis 1:28) God commanded Adam to *know* his wife Eve in the strictest biblical sense of the word and to have children and to bring them up in the ways of the Lord. This is the way that God's people were to spread across the entire earth, subduing it. The Fall did not negate this mandate. Repeatedly, Scripture confirms the primacy of teaching our children the ways of the Lord. God calls us to make disciples, starting with our own children.

Here we see the unique role and ability of the woman to steward life, to give birth, and her unique bend toward nurturing. Secular culture overlooks this unique capacity to *mother* as women desperately seek to usurp what was not given to them in the first place. What an amazing thing, to mother, care for, and nurture life from its very inception at conception. I stand in awe of this capacity as should any godly society, holding the desire to mother as a truly sacred desire.

Please do not hear me say that a woman cannot and should not work or pursue a profession. By all means, I merely proclaim the sacred call of *motherhood* and the disparate ordained roles for men and women.

3. **Marriage has the mission of proclaiming the Gospel.**

As I love my wife as Christ loved the church and she submits to me as to the Lord, we display the Gospel. The Bible frequently refers to the Church as the Bride of Christ and its relationship with Jesus as a marriage. Betrayal of Jesus is likewise discussed in terms of adultery, spiritual adultery. The final consummation of the Church, her reunion with Jesus in heaven at His Second Coming, is even described as a marriage feast. (Revelation 19)

Based upon these two purposes, is it any wonder that Satan and the world have continuously sought to undermine the sanctity of marriage, driving a wedge between man and woman? They often rely upon factors from point number one in distorting and perverting godly roles to exacerbate the conflagration between the two sexes.

Our nation continuously drives further from the godly standard taking our army with it. Like the populace, our soldiers possess intense confusion over issues of masculinity, femininity, and sexuality. This resonates in their actions and vast affliction.

Gender Friction

In one sense, all sin is the same. It all condemns equally. John says that, "Everyone who makes a practice of sinning also practices lawlessness; sin is lawlessness." (1 John 3:4) When it comes to eternal condemnation, all sin equally condemns men to an eternity apart from God.

In another sense, there are degrees and types of sin, some of which merit discussion of degrees of punishment while in eternity apart from God. Sexual sin merits special consideration. Paul writes to the church in Corinth, a church suffering due to accepted and pervasive sexual sin,

> *Flee from sexual immorality. Every other sin a person commits is outside the body, but the sexually immoral person sins against his own body. Or do you not know that your body is a temple of the Holy Spirit within you, whom you have from God? You are not your own, for you were bought with a price. So glorify God in your body.*
> *1 Corinthians 6:18-20*

Sexual sin is unique in that it sins against the body, the temple of the Holy Spirit.

Consider just a few of the general afflictions which departure from a godly sexual ethic yields. Countless women suffer the violence and trauma of rape and sexual assault. An entire generation of men presently come of age addicted to pornography, incapable of true intimacy with a woman. Children born out of wedlock and raised by single mothers, over 40% of current births, suffer in life in comparison to those raised in a two-parent household.[98] We have murdered more than a generation of Americans in the womb, upward of 60 million since 1973, a direct attempt to mitigate the consequences of sexual sin. Millions of Americans suffer with a sexually transmitted disease. These are but a few of the ramifications of sexual sin,

---

[98]Michelle Castillo, "Almost half of first babies in U.S. Born to unwed mothers," *CBS News*, accessed October 14, 2016, http://www.cbsnews.com/news/almost-half-of-first-babies-in-us-born-to-unwed-mothers/.

and as they resonate in society they resonate throughout the ranks.

General Ray Odierno, then the Chief of Staff of the Army, declared in testimony before the Senate Armed Services Committee in June 2013 that,

> *"combating sexual assault and sexual harassment within the ranks is our number one priority."*[99]

Call me pedantic, but I always thought that closing with and destroying the enemy ought to be the number one priority of any Army. General Odierno had no choice. The civilian leadership was threatening to remove the authority of the military to police itself under the UCMJ, so pervasive had rape and sexual assault become in the ranks.

Illusion of Gender Equality

I'm no Rambo, but I have never met a woman I couldn't kill with my bare hands. That is not a statement of bigotry or hate, but a statement of fact that bears relevance to several discussions. One of the primary manifestations of wickedness in the hearts of men is the suppression of women. The military in many ways unwittingly encourages sexual violence against women by bending to our civilian masters in refusing to acknowledge the fact that gender equality is an illusion, a charade, and harmful one at that.

Gender equality only exists as allowed by a system. Absent a permissive society, in the presence of anarchy, women are decidedly vulnerable due to their weaker bodies and kinder natures. Do exceptions exists? Undoubtedly, and women are as capable of brutality as any man. In general, men possess a corner on the market for the application of brute force and brutality. Almost all violent crime is committed by young men. Almost all domestic violence is committed by men. Men, unrestrained, excel in the oppression of women and the application of violence. This is a fact borne out by history and declared by God in the Garden. (Genesis 3:16)

The key notion is restraint, and it is the Gospel of Jesus Christ and the previously mentioned idea of common grace that restrains the brutality of men and provides the conditions for equality. Jesus was the greatest proponent and protector of women. In the patriarchal 1st century Jewish

---

[99] Accessed May 23, 2016, https://www.army.mil/article/104753.

culture, women were subservient, second class citizens living completely at the mercy of their male overlords.

This is also the case in virtually every other society since, other than those based upon a Judeo-Christian heritage. Can you name a single matriarchal society? I can't. They don't exist; they never have. Atheistic, Buddhist, or Hindu Eastern societies are all decidedly patriarchal. Never mind that every single Muslim society openly oppresses women, many in an extremely brutal fashion. No, it is the common grace of God through the Gospel of Jesus Christ that protects women from the sin of men.

I have an Aunt, an elderly lady with a decidedly liberal outlook on all matters. In the ongoing national debate concerning bathroom use and allowing men who identify as women to use female bathrooms and changing facilities, she made the comment to me that she didn't need my protection in a bathroom. She could take care of herself.

During the 2016 Miss USA pageant, the eventual winner, Miss District of Columbia who also happens to be a reserve Army officer, was asked about the Pentagon's recent decision to open ground combat positions to women. Without hesitation, she declared her enthusiastic support, "We are just as tough as men."[100]

Both of these women live in fantasy world. My Aunt is a little old lady and any grown man who followed her into the bathroom could literally do whatever he wanted, were it not for the intervention of…another man. The veracity of Miss USA's statement depends on what you mean by tough. My wife is one of the toughest people I know. She is a fearless mother, tireless and strong. In fairness, I would not want her next to me in a gunfight. Miss USA's statement rings true while she's wearing a glittery dress, replete with tiara and a bouquet of roses. Put her under a rucksack confronted with a bevy of grown men trying to kill her and her comrades, and the charade loses its luster.

Refusing to acknowledge the inherent godly differences and that a system is necessary endangers women and potentially the mission. Removing the veil of Christ's protection and provision for women places them at risk, yet we cannot acknowledge that lest we offend the sensibilities of the secular masters of this nation.

---

[100] Penny Starr "New Miss USA and Army Vet: 'We are Just as Tough as Men'," *CNS News*, accessed October 14, 2016, http://cnsnews.com/blog/penny-starr/new-miss-usa-and-army-vet-we-are-just-tough-men.

The Rhino and the Butterfly

Peter writes,

> *Likewise, husbands, live with your wives in an understanding way, showing honor to the woman as the weaker vessel...*
> *1 Peter 3:7*

As much as concepts of submission offend so readily, so too does the idea of weakness. Many seethe over the connotation, deny its truth. Consider another angle. Consider weakness in terms of fragility, as in the fragility of a ceramic vase—weak, fragile, beautiful really, delicate, and valuable. Consider a ceramic vase versus a steel pot, the man, or perhaps a butterfly versus a rhinoceros. Some situations require the attributes of a rhino. Would you really send a butterfly to do the work of the beast?

A recent deployment aboard an Air Force C-17 illustrated this idea nicely. The assistant loadmaster was a young lady, though it took me a minute to realize it. She sported a short, boyish haircut and walked and acted like a man, sort of. It took me a minute to realize that this was, in fact, a female. Once I did, it was extremely obvious that this was a female acting like a male. Now, whether she was 'trans' or not, whatever that actually means, I don't know. Maybe she was just a boyish female. Either way, her femininity was obvious despite her best attempts to portray masculinity.

It became even more obvious when she had to *do* things. The loadmaster on a C-17 is responsible for all of the cargo, to ensure it is loaded correctly and safely. At one point, the head loadmaster directed the young woman to secure a pile of plastic boxes with a cargo strap. I stood out of the way and watched this young lady fumble with the industrial strength cargo straps for several minutes with no success before needing the assistance of the loadmaster, a man. She simply did not possess the strength and leverage to make the straps work.

Shortly before takeoff, the loadmaster directed the young lady to close the door to the aircraft and yet again, she could not complete the task, lacking the strength and leverage. After several failed attempts, she once more required the assistance of the loadmaster, a man. He walked over, casually threw his weight into it, and slammed it shut.

This young lady literally could not generate the torque and leverage that the man could though he was no bastion of masculinity. She actually appeared to be in much better shape. He was a middle-aged, slightly overweight, E-7 with a beer belly. Yet, he could generate the brute force that this young lady could not, and no matter how much she wanted to look like a man the instant brute force became necessary, her femininity became intensely obvious.

This is not to impugn this young lady. I maintain great respect for her service and the fact that she needed assistance is no issue. Certainly there are men who might require assistance with these mundane tasks though I've never observed that. What I seek to highlight is that to ignore the distinction between men and women is a most foolish thing to do. She was a butterfly and that fact was never more obvious than when she was required to do that which is expected of the rhino.

In this case, the inability to generate brute force did not prove decisive. However, I can think of numerous situations where this ability might just be the difference between life and death. Situations exist where the restraint of the Gospel ebbs, where the smoothing effects of civilization and society wane, where the ability to generate and respond to brute force and brutality might just be the difference between victory and defeat. I'm thinking specifically of combat, definitively ground combat.

The Obama administration recently opened all positions in the military to women despite the fact that a Marine test of infantrymen versus integrated infantry proved beyond a shadow of a doubt that women do not perform well as infantrymen.[101] Go figure. We don't need tests to demonstrate this. Anyone who has served, including most women, will attest that this is an endeavor fraught with peril that will ultimately cost lives. Gender equality is good as long as we understand its precise nature and how precarious it may be.

Further Fallout

I anticipate that, perhaps apart from the actual Gospel presentations, this section on biblical sexuality will likely generate the most condemnation. Our nation is so far down this path as is our military that I'm just not sure that it would even be possible to revert.

In light of the inevitable cries of sexism and bigotry, allow me to say that I love, respect, and cherish women. I am married with three daughters of my own whom I love more than life itself. My military experience has

---

[101] Jim Michaels, "Marine study finds all-male infantry units outperformed teams with women," *USA Today*, accessed October 14, 2016, http://www.usatoday.com/story/news/nation/2015/09/10/marine-study-finds-all-male-infantry-units-outperformed-teams-women/71971416/.

offered me some unique insight into gender inclusion. The unit I served for so many years never had women below the Brigade level though it has since been integrated. At one point, we literally had 13 women in an organization of over 3,000.

When I got sent to the Division, I found myself surrounded by women. I had female commanders, female 1SG's, female lieutenants and a litany of female soldiers, and many of these soldiers were intensely professional and committed. I have made a few observations.

More than a few of the young female soldiers were single mothers who had replaced their child's absentee father with the Army. They had literally replaced the stability and support that the father is supposed to provide with the stability and support of the Army, and again many of them were great soldiers for whom I have tremendous respect.

As the rear detachment Brigade Commander, my S3, the officer in charge of running the entire Brigade, a position normally reserved for a major, was a female captain. She was one of the best officers I have worked with. Intelligent, physically fit, and motivated—I rated her as #1 among the nearly 30 officers I rated in her grade. I would proudly serve with her again. As much as we had professional and committed female soldiers, we also had plenty of dirt-bag female soldiers, overweight and unmotivated, just like some of their male counterparts.

A soldier is a soldier, or so goes the mantra. As Guttman observed in her work, *The Kinder, Gentler Military*, when you remove all the other differences and treat everyone the same, sexual difference is the only thing you notice.[102] As much as attempting to treat men and women the same could potentially damage the mission, its potential to damage women is even greater.

I went to some Army training once called *Got Your Back* training. It was hosted by a young civilian man and woman who used crude language and talked about crude things in an effort to be hip and relevant, I guess. The whole point of the training was to teach the young soldiers how to safely 'hook-up'. How do they determine when the 'hook-up' transitions to the assault and what to do to keep the 'hook-up' from turning into an assault?

At one point they asked, "What is the number one thing you can give a woman to ensure a safe hook-up?" expecting "a condom" as the answer.

---

[102] Stephanie Guttman, *The Kinder, Gentler Military*, (New York: Scribner, 2000), 61.

"A wedding ring," I thought to myself.

A couple of points here. First, this type of training and mindset makes perfect secular sense. They will go to the club anyway. They will 'hook-up' anyway. Why not teach them how to do it safely? This is the same rationale parents use in providing their young daughters with birth control. I still remember a conversation between my oldest daughter, Ami, and myself a number of years ago.

"You know my friend is on birth control."

"Good for her."

"What would you say if I asked to be on birth control?"

"We would say, 'not in this life'."

"But what if I got pregnant?"

"Then don't have sex."

"You wouldn't give it to me?!" was my daughter's incredulous reply. Biblical sexuality ran in complete contrast to the secular views of her friends and their parents.

I stayed after the *Got Your Back* training to rebuke the instructors who offered to me that abstinence education has been proven largely ineffective. The issue is the context. We are at the point whereby we assume sin in our soldiers. I ask another question, "What if they aren't there in the first place?"

Proposed forced equality, the denial of complementarianism, exacerbates the issue and contributes to the problem of sexual assault and rape. The Army has stayed the course in attempting to treat males and females the same, something that was entirely shocking to me upon my reemergence into the force. In the field, they even share sleeping tents. There is no female tent. I found this highly unusual. The Modern Army Combatives Program (MACP) is another area of equality that spotlights the unavoidable.

In Combatives, there are no males or females, only soldiers. This means any male soldier could find himself grappling with a female soldier. Combatives looks like a cross between wrestling and Ju-Jitsu and places the opponents in what could be considered very intimate positions. Bodies are pressed against bodies. Legs are wrapped around waists. Again, call me antiquated, but I found this to be very inappropriate. As we instituted a robust Combatives program, I frequently found myself confronted with the prospect of engaging with a female. (I would combat this by pretending to be too tired at that point and needing to sit out for a round or two)

Aside from the inappropriate contact, one engagement I witnessed highlighted a more troubling aspect. I had a young lieutenant built like Mike Tyson, thick and powerful, taking on a young female officer who was built like a typical female, small and frail by comparison. It was a joke. The male kind of toyed with her a bit, somewhat bemused by the situation and more than a little uncomfortable, and then at some point, he kind of grabbed her as gently as he could, and pushed her to her back and achieved the dominant mount position. She even uttered out a little cry as took her down. Now, he did it as gently as he could. To me, it looked like domestic violence practice.

As men tend to sin by oppressing women, we actually denigrate women by placing them on the same level as men. We, men, should actually elevate the position of women. My wife and I teach all six of our sons that you "treat a girl like a flower." If they hit one another then okay, "no blood, no foul," but if they hit a girl, my granddaughter for instance, I come down on them with the full wrath and fury of the Lord, Almighty.

Return with me once more to the Garden, to creation. Consider that God made everything—the heavens and the stars, the sun and the moon, the earth and the sky, the beasts of the field, the fish of the waters. God made everything by His spoken word, and then He brought forth man and put Him in the Garden to work it, to have dominion over it. He taught man His statutes, and only when all of creation was prepared and ready, poised, He brought forth woman. Adam then exclaimed, "This one at last, is bone of my bone, and flesh of my flesh." The Bible adamantly speaks to the elevation of women by men, lest they be oppressed by those very same men.

We see this in the ranks. Consider a typical sexual assault. A young male soldier lives in the barracks right next door to a female. He went through basic training alongside her. They shared a tent together. She is just 'one of the guys'. He fights her in Combatives as again, she is just 'one of the guys', nothing special, no different than his other friends, and then they go to the party and get drunk together. As they've been encouraged to 'hook-up' and he is feeling the need, he has his way with her, whether she is willing or not as they end up in his barracks room or a buddy's house downtown.

Instead of viewing her as something special, something to be cherished and respected, he has been taught by programs absent a biblical foundation of complementarianism that she is really no different from him aside from the fact that she possesses a vagina. Men take what they want from other men all the time, by force if necessary. Why not do the same with her?

I return to my daughters. When I think of them and their prospective husbands, I don't want for them a man who treats them with equality. I don't want a husband for them who treats them the same as he treats his buddies. My desire is that a man comes along and treat them as a princess, that he cherish them and hold them in the highest regard, that he sees them as the weaker vessel, the ceramic vase, the butterfly that they are and conducts himself accordingly—gentle, honorably, with respect.

In striving for what we designate as equality, our nation and our Army endangers women, actually placing them in a lesser position than God intended. And, as they'll soon be sharing a bathroom, shower facility, and locker room if our secular overlords have their way, it will only get worse.

Sexual harassment is merely a tamed down version of these same issues with other exacerbating factors such as rank and authority. The proliferation of domestic violence, both in the military and society in general, also illustrates these points. Yes, women commit domestic violence and sexual harassment/assault on occasion but it is rare, despite the best attempts by the powers that be to convince themselves otherwise. The vast majority of sexual sin involves a man abusing a woman in some way.

Speaking to resources, the Army has developed an entire cottage industry to address the problem of sexual sin from a secular vantage. We've flooded the ranks with programs and counselors and specialists trained to deal with sexual assault, civilian and military. We require mandatory training and classes. Posters, displays, and banners decorate the headquarters imploring of the soldiers, "Not in my Squad" and "I.A.M. Strong!" Has any of this made a difference?

I've noted essentially two types of sexual assault. There is the drunken hook-up that went too far, almost always a young man on a young woman and then there is the pedophile, the predator. Unfortunately, predators do roam the ranks, a minuscule percentage, but there just the same. These are the child rapists or otherwise deviant perverts. No program will deter the predator; they just need an opportunity.

Unfortunately, the SHARP program, the Army's flagship in the war on sexual sin, does very little to actually address the roots of any sexual issue within the ranks.

Clarification

Before proceeding, I am led to clarify a point. Lest you misunderstand, I am advocating for the distinction of women, the opposite of degradation. I am an advocate for women in the service. I have served alongside amazingly competent women, officers and enlisted, every bit the warrior and patriot of their male counterparts. However, perhaps we can do it better, honor both the unique nature of men and women while simultaneously utilizing their requisite skills for the defense of the nation.

Solution

It is here that my assessment falls admittedly short. The most obvious answer is, "Repent!" Collectively, as a nation, the only solution is revival, to abandon the secular slide and turn once more to the ways of the Lord. Apart from a dramatic supernatural intervention in this fashion, the question must be framed appropriately, an exceedingly difficult venture.

"What do we do?" becomes,

"How do we generate godly sexual behavior while still clinging to and teaching a godless sexual ethic?" which specifically translate to,

"How do we teach our young men to respect and cherish young women while still teaching them that they are the same, minus the presence of a penis or a vagina?"

We should immediately abandon the farce that men are the same as women. Allow men to be men and women to be women. Re-segregate basic training. Re-segregate billeting. Re-segregate sleeping quarters. Re-segregate combatives training. Inject a manner of separation and quit trying to teach men and women that they are the same. Quite the opposite, we ought to honor and celebrate our God-given sexuality, male and female, instead of foolishly pretending that once a person slaps on a uniform, they somehow become asexual.

Second, we should at least teach a godly ethic as an option. If we continue with *Got Your Back* training, at least tell them the truth, that there is another option, another way.

Third, we should seriously review and consider the introduction of women into ground combat and their inevitable inclusion in the Selective Service. Can women fight? Of course. I think of Stalingrad and Kobani, yet these were unique circumstances for a desperate time. Description is not identical to prescription, and we need to be able to differentiate. A better question that we must ask is "ought a woman to fight?" Should we intentionally send our women to do the dirty business of ground combat?

As I know that these things will likely never happen—we've drifted too far—I return to my original plea for America.

"Repent."

27. Affliction of a Godless Army

One only needs to serve for a short period of time before becoming intensely aware that our Army is an afflicted army. You may disagree with me concerning the source and that's okay, but one may not deny the affliction minus some extreme intellectual dishonesty. Returning to the divisional Army after 13 years, nothing could prepare me for the reality of our soldier's afflictions.

Aside from sexual sin and a vast misunderstanding of masculinity and femininity, numerous other afflictions scourge our armed forces today. We'll address but a few.

From the Top

I cannot address the sin of the rank and file without first addressing the sins of generals. When the soldier partakes of the maddening wine of the whore of Babylon, the fallout might resonate across his unit or his family, but when a senior officer transgresses, the fallout can resonate across the legions.

In recent years, the United States military finds itself wrestling a surge in senior officer misconduct. From 2011 to 2014, the number of Army officers disciplined for misconduct has tripled.[103] Why wouldn't it? As godlessness pervades and the common grace of the Gospel is suppressed, unrighteousness penetrates to every echelon. Godless generals pursue godless affairs as does any fallible man.

The most widely-known account is the tragic fall of General Petraeus

---

[103] Mark Wilson, "Number of US army officers fired for misconduct tripled in 3 yrs—report," Reuters, accessed 25 September 2017 at https://www.rt.com/usa/army-misconduct-soldiers-rise-280/.

who got caught in 2012 in an adulterous affair with his biographer, Paula Broadwell. His fall stunned as Petraeus was a star, one of the most successful officers ever produced. "This is a man who has never failed at anything."[104] But fail he did, losing his job, his reputation, and damaging the nation's war effort in the process.

He's not alone. There's the Major General who got exposed by an anonymous tip. Leading a licentious, "swingers" lifestyle, visiting sex clubs, having sex with multiple partners, for three years he betrayed his family and his profession. Then there's the general who had a three-year affair with a junior officer.

A Brigadier General gets fired for assaulting his mistress. Another is busted using his government credit card at strip clubs. Still another general sexually harasses subordinates in a sex-and-bribery scandal.

One drinks on duty. Another accepts illegal and expensive gifts from foreigners. Another treats himself and his wife to a $750-a-night Caribbean hotel at the taxpayer's expense. Another general takes numerous unauthorized, personal trips in government aircraft, also on the taxpayer's dime.

Another general sends repeated, racy texts to a junior enlisted soldier's wife. Still another makes off-color jokes in an email to other generals about masturbating to the 'hot' visiting congresswoman.

Epitomizing the affliction, Major General John Rossi committed suicide becoming the highest ranking officer to ever take his life. A West Point graduate who was set to be promoted to Lieutenant General and assume command of the Army's space and missile defense command, Rossi took his own life and left behind his wife and three children.

No senior officers get fired for incompetence. The competition is fierce and most senior officers are driven and highly competent individuals. Senior officers generally transgress in one of three areas:

- *zipper* issues,
- *bottle* issues,
- *money* issues.

A cursory review of the individual accounts reveals every single one of these, especially sexual transgression. This invites some uncomfortable

---

[104]Michael Pearson, "The Petraeus affair: A lot more than sex," CNN.com, accessed 25 September 2017 at http://www.cnn.com/2012/11/12/us/petraeus-cia-resignation/.

question. How does this happen? Where is the accountability?

At some point, the institution begins construction of the cult of personality as we build leaders into hyper-actualized images of their true selves. As you progress in rank, people begin to want to do things for you, edging into moral and ethical gray areas. Senior officers become increasingly isolated and insulated, with fewer peers, surrounded by affirming subordinates.

Did no one notice General Petraeus spending so much time with a younger woman, not his wife? Did no one have the wherewithal to pull him aside and ask what was going on?

These are fallible men, no different in their capacity for sin than any other men. Yet, the military puts them on a pedestal, gives them access to things that no other soldiers has, and removes accountability. Then, we are astonished when they fall.

In dealing with this issue, the Army is putting together new mental health, counseling and career management programs "to shape stronger, more ethical leaders."[105] These are mere secular band-aids on a gaping spiritual wound. This is a spiritual issue, the permeation of godlessness into the ranks of our senior leaders.

The force has no choice but to follow suit.

The Scourge

The ever-present demon of chemical addiction exacerbates the plight of soldiers. Alcohol and drug abuse scars the souls of these young men and women. With ease, I envision one of my young soldiers in his barracks room all weekend, alone with Satan himself. Satan, always whispering,

"Do it, there's no hope. Do it. It'll make it better."

"I don't want to."

"It's the right thing."

"I'm afraid."

"They'll call you brave."

"I don't want to."

---

[105]Lolita C. Baldor, "Army looks for new ways to address misbehaving generals," AP, accessed 25 September 2017 at https://www.yahoo.com/news/army-looks-ways-address-misbehaving-generals-092224860--politics.html.

"Have another drink."

I mourned at the consideration, that the blackness in a young man's heart was further bolstered by the consumption of alcohol to the point whereby the demonic could convince him to cinch his belt around his neck, secure it over the door jam, and literally sit to his death. I still mourn at the sheer emptiness that led to such an act, the spiritual bankruptcy enabled by chemical enslavement.

I felt compelled to write my inaugural work, *Scourge: Confronting the Global Issue of Addiction*, before ever setting foot in the Division. I noticed, via the foster care system, the surprising prevalence of alcohol and drug abuse. We live in the Bible Belt yet, every single foster child we'd sheltered over the course of eight years—numbering nearly 30—dealt with the impacts of addiction. Many were taken into custody due to the sins of their birth parents in terms of addiction. Some birth mothers afflicted their children in the womb, consuming drugs and alcohol while pregnant. My seven-year-old son, our first foster child nearly seven years ago, suffers from a litany of physical ailments due to his mother's crack use during pregnancy. Physical and emotional abuse frequently accompany addiction. My fifteen-year-old son spent time in a crack house with his biological parents some years ago. It's a wonder he can even function.

I should not have been surprised to find the same issue in the Army. I'm not even sure how to adequately capture the extent of the issue, how to adequately do it justice. I recall reading about the 'Hollow Army' of the 1970's and the rampant drug abuse that permeated the ranks.

Consider that the Army is sending kids as young as 20 years old to in-patient rehabilitation. Consider that the vast majority of rapes and sexual assaults occur within the context of binge drinking. Cocaine, marijuana, spice, prescription drugs—they are all present and prevalent. The statistics paint a surprisingly bleak outlook concerning the widespread abuse of chemicals in the ranks and as I said, it normally accompanies other afflictions, weakening minds and eroding the will. People will simply do things under the influence that they might not otherwise do.

Three young soldiers, two males and a female, return to the barracks in a drunken stupor and have a sexual triste. The next morning the female cries, 'Rape' while the young men cry, 'Consensual'. Either way, lives are ruined. The van ride to Fort Leavenworth must've seemed to take an eternity for my young sergeant convicted of raping a junior soldier. He was drunk at the time as was she. Another young soldier, on more psychotropic medications than any man should be due to a laundry list of behavioral health diagnoses, weeps feebly at the slightest demand. He is literally, incapable of working.

A young sergeant shoots up a local nightclub. Cocaine and alcohol are involved.

The deviant creativity of the enemy ensures soldiers will remain one step ahead of the authorities. As fast as the army responds to one type of abuse, someone invents another. Recently, soldiers started spiking their 'vape' e-cigs with all manner of chemicals in pursuit of the high, undetectable to any current methods.

The moral bankruptcy in the hearts of these soldiers drives their pursuit of worldly satisfaction. Apart from Christ, godless soldiers will seek the high in any way including chemical abuse. The increasing godlessness of our soldiers should cause great concern as the scourge of addiction will likely continue to grieve the ranks for the foreseeable future.

Suicide

Few things consume a unit like a suicide. My brigade had two in a matter of months. On one particular weekend, a young soldier full of heartache and alcohol hung himself in his barracks room. He and his girlfriend were having some significant relationship issues. A few months later, another young soldier, the one we spoke of above, hung himself with his belt. Hours before his death, he posted a picture on social media of him in his barracks room, alone…with a bottle of liquor. Shockwaves roiled throughout the Brigade. Neither young man had previously displayed overt suicidal ideations.

Thankfully, they came from separate battalions, but in the immediate aftermath and for days and weeks following, the units were consumed. The chain-of-command was focused entirely, as it should've been, as it had to be, upon the care of the family and the unit. We sent teams to funerals, executed memorial ceremonies, offered condolences to families and supported them any way we could. More than that, we tore ourselves apart, seeking answers that never presented themselves. How could we have prevented this? What more could we have done?

We were asking the wrong questions.

As suicide proliferates throughout the active ranks, it likewise afflicts our nation's veterans. A popular narrative claims that 22 veterans commit suicide every day which translates to one roughly every 65 minutes. 22 suicides a day—Politicians regurgitate it, veteran's groups made it a banner, and sympathetic citizens demand answers. Now, even one suicide is too many. Yet I wondered, if this is accurate, then this is an astonishing number!

The statistic, 22 a day, is based upon the Veteran's Administration 2012 Suicide Data Report which surveyed death statistics from 1999 to 2011 across 21 states and then extrapolated for the population. Here is the issue, the span of the survey calls the data into great question as the researchers themselves urge caution. "It is recommended that the estimated number of veterans be interpreted with caution due to the use of data from a sample of states and existing evidence of uncertainty in veteran identifiers on U.S. death certificates."[106] Furthermore, the average age of the suicide victim in the study was 60 years old further undermining the narrative of afflicted Afghanistan and Iraq veterans that so many have come to depend upon over the years. A more recent and comprehensive survey yields that roughly one veteran actually commits suicide each day.[107] Again though, even one is too many as suicide rates among the general population, particularly the young, have been steadily rising over the last decade.[108]

As a young officer, I scarcely recall a suicide, not a single one. What has given rise to this phenomenon among the active and veteran ranks? Do the current wars truly afflict our soldiers to this point of desperation and affliction whereby they view suicide as their only option or source of relief? Perhaps, though increasingly most suicide victims in the active ranks have never even deployed.[109] How do we reconcile that?

A number of factors contribute to suicide. However, we can trace its origins to the singular condition previously noted, the spiritual bankruptcy resident in the hearts of men apart from the Gospel of Jesus Christ. The proliferation of the unchurched in the ranks effectively sets the condition for numerous abominable practices, including suicide.

---

[106] Michelle Ye Hee Lee, "The missing context behind the widely cited statistic that there are 22 veteran suicides a day," *The Washingon Post*, accessed May 24, 2016, https://www.washingtonpost.com/news/fact-checker/wp/2015/02/04/the-missing-context-behind-a-widely-cited-statistic-that-there-are-22-veteran-suicides-a-day/.

[107] Stacy Bare, "The Truth About 22 Veteran Suicides a Day," *Task and Purpose*, accessed October 14, 2016, http://taskandpurpose.com/truth-22-veteran-suicides-day/.

[108] American Foundation for Suicide Prevention, accessed October 14, 2016, https://afsp.org/about-suicide/suicide-statistics/.

[109] James Dao and Andrew W. Lehren, "Baffling Rise in Suicides Plagues the U.S. Military," *The New York Times*, accessed July 12, 2016, http://www.nytimes.com/2013/05/16/us/baffling-rise-in-suicides-plagues-us-military.html.

As the Gospel is suppressed, men lose value. Secular, agnostic, or even atheistic thought systems deny the inherent value of men as the *Imago Dei*. Regressing to ultimately evolutionary constructs, men become merely the latest and most adapted of all creatures. Men possess no intrinsic value and life has no intrinsic worth. A proper understanding of the Image of God produces in a man's heart a respect and value of all human life. All men's lives hold sacred value, including their own and as such, life must not to be taken lightly.

Along with an understanding of the sacred value of life, with the Gospel comes hope. No matter how desperate a situation, the believer lives with a hope not found in himself, rather a hope found in God. I have the hope of things not yet seen, the glory of a future spent in eternity with the Lord our God. As a believer, I am subsequently empowered by the Holy Spirit to accomplish that to which He has called me. I may live well, overcoming all oppression and persecution…or I may die well. Either way, I have a hope in my soul that cannot be overcome and a Savior that loves me without end.

It is hard to say and to those who have been affected by the suicide of a loved one or family member, I apologize profusely for the following statement, but it must be said. Suicide is an intensely selfish act. The victim is normally completely absorbed by the affliction of their existence, completely hopeless and ill-equipped to deal with the afflictions of life. The Christian life calls the believer to the exact opposite, to be consumed first with God and then with the life and welfare of others. It is hard to imagine a believer focusing enough on himself to consider suicide.

But it happens. Non-believers are not the only ones who commit suicide. I knew a great chaplain once, a great man of God, a man who loved the Lord and his family. He took his own life. He had been caught up in sin and the devil talked him into it. The tragedy of suicide emanates from its irreversibility.

Now, these are general themes, the moral and spiritual bankruptcy apart from Christ that generates the conditions whereby soldiers consider suicide a feasible, acceptable, and suitable course of action. Vast and infinite mitigating and modifying circumstances exist. Yet, a direct correlation exists between the proliferation of the unchurched and the increase in suicides and suicidal ideations both in our society and in our nation's Army.

I am led to address the heretical belief that suicide automatically condemns a man to hell. This stems from a fundamental misunderstanding of salvation itself. Salvation is of the Lord, nothing you do merits salvation, and once you are sealed unto God, you are sealed unto eternity. If you could lose your salvation, you would. Suicide is a sin, no doubt, and as you are dead, you will not be able to seek forgiveness yet, the only sin that condemns to hell is the sin of unbelief. If one believes and commits suicide, he may yet have to account for taking his own precious life, but he will not be condemned. "There is therefore now no condemnation for those who are in Christ, Jesus." (Rom. 8:1)

And still the Army calls upon commanders to account for and deal with this plague at the expense of preparations for war.

Atrocity

The delicate strife of warfare offers much in the way of opportunity for evil men. Such was the case on a sweltering Sunday in Yusufiyah, just west of Al-Mahmudiyah, right in the heart of the Triangle of Death as five American soldiers descended upon the Iraqi farmhouse.

The proud 1st Battalion of the 502nd Infantry Regiment, 2nd Brigade Combat Team, 101st Airborne Division (Air Assault) had its hands full on this deployment. At about the halfway point of their tour, they had already lost over 30 men and would go on to lose a total of 67, battling insurgents and IEDs as they struggled to impose order.

On March 12th, 2006, not far from where Worrell and Weeks would give their lives, five soldiers abandoned their post at a vehicle checkpoint. The junior member, Private First Class Steven Green led them to a nearby farmhouse they had previously searched and first noticed 14-year-old Abeer Qassim Hamza al-Janabi.

In broad daylight, they kicked in the door, surprising the family. They quickly sequestered Abeer into a room by herself where two of them raped her while Green murdered her parents and younger sister in another room. Green then raped Abeer and shot her in the head before setting her on fire.

All five returned to the checkpoint and went back to their business. When the word of the crime came out, the soldiers blamed it on local Sunni insurgents. Sergeant Anthony Yribe was one of the first at the scene of the crime. From there, he went to the checkpoint where he found Green and the others. Almost immediately, Green confessed. SGT Yribe kept it quiet.

On June 16, in an apparent revenge killing, Sunni insurgents overran a nearby checkpoint, killing Specialist David Babineau. Privates First Class Thomas Tucker and Kristian Menchaca were captured, brutally tortured, and killed. Upon hearing this, SGT Yribe confessed to Private First Class Justin Watt what had happened. Watt revealed the crime during a psychological health screening.

These were American soldiers, young men with a future, young men thrust into the heart of darkness, not so different from any young man today. It's not like these were serial killers, though prior to Watt revealing the crime, Green was discharged with an anti-social personality disorder.

Atrocity has always been a part of war. All armies have committed atrocity at some point as the enemy and evil men seek to capitalize on moral latitude and the freedom of maneuver offered by combat. Godlessness contributes to atrocity, either in a localized or pervasive manner dependent upon the extent of the godlessness. The Judeo-Christian ethic, the proliferation of the common grace of the Gospel, quells the potential for and toleration of, atrocity.

Notice the godlessness ethic evident at Mahmudiyah. The senior member of the group, a sergeant, stood impotent as the group openly discussed their plan while consuming alcohol. A sergeant stood by and listened as his soldiers drank alcohol on duty and plotted a rape and murder! He remained silent as the junior member of the group, a known malingerer, led the group to abandon their post and commit this crime and then participated himself.

Think back to the moral courage exhibited by Master Sergeant Roddie Edmonds as he confronted the Nazi prison warden. Consider the moral courage of Warrant Officer Hugh Thompson at My Lai. Both stood to lose much, possibly their lives, but chose to stand between the innocent and evil. Jesus codifies this mindset. Greater love hath no man…Instead, a man with a godless coward's heart stood silent as an innocent girl suffered at the hands of evil men.

Godless devaluation drove this crime. After his trial, Green was quoted as saying, "I didn't think of Iraqis as humans."[110] After several of his friends were gunned down at a checkpoint, Green's hatred intensified. "There's not a word that would describe how much I hated these people."[111] Green and the others clearly thought lesser of the Iraqis.

When they saw young Abeer, they didn't see a little girl with hopes and

---

[110] "'I didn't think of Iraqis as humans,' says U.S. soldier who raped 14-year-old girl before killing her and her family," *Daily Mail,* December 21, 2010, accessed September 17, 2017 at http://www.dailymail.co.uk/news/article-1340207/I-didnt-think-Iraqis-humans-says-U-S-soldier-raped-14-year-old-girl-killing-her-family.html?ito=feeds-newsxml.

[111] Ibid.

dreams. They didn't see the Image of God, a fellow human. They saw an inanimate object, a potential source of pleasure and satiation. Their godlessness resonated in devaluation and stratification. As Green himself later admitted, he wouldn't be much of a human to think these things were okay, in retrospect.

Armies throughout history have fomented devaluation as a means to encourage killing. The American army has fought Japs, Krauts, Nips, Gooks, and Ragheads. Devaluation makes it easier to take lives, but at what expense? As we've seen the rightly motivated soldier doesn't require devaluation to fight. Institutional devaluation coupled with pervasive godlessness ripens the fields for the slaughter.

What good is Christ if a so-called Christian army commits atrocity?

First, I'll argue that atrocity from armies grounded in the Judeo-Christian ethic are normally an aberration. In fact, I've seen the opposite effect as armies place their own soldiers at great risk in an effort to protect innocent lives. Godly armies don't use non-combatants as human shields. Godly armies, as institutions, do not commit atrocity or sanction atrocity. Again, aberrations exist, but the second argument I'll make speaks to accountability.

Godly armies hold those who do commit atrocity accountable. From My Lai to Abu Ghraib, soldiers who commit war crimes are called to account. Yes, they are called to account imperfectly. Sometimes the system hammers junior soldiers while more senior soldiers and officers slide by. Sometimes the punishment doesn't fit the crime or maybe the whistle-blower is persecuted. See Warrant Officer Thompson. Sometimes the guilty go unpunished or higher seems willfully blind to an incident. The system is far from perfect but what is the intent?

The intent, the heart of a godly army, is to hold its soldiers to a higher standard and when they transgress, to hold them accountable. This is unique to the godly army.

SGT Yribe plea-bargained for a discharge in exchange for his testimony. PFC Howard, the lookout, served 27 months and was discharged. Barker, Cortez, and Spielman remain incarcerated at the United States Army Disciplinary Barracks at Fort Leavenworth, Kansas. Green was sentenced to life without parole when the jury could not come to a unanimous death penalty recommendation.

In February 2014, Green hung himself to death.

The spread of godlessness in our army should cause great concern as we put soldiers into morally ambiguous situations with a gun in their hand.

Turning the Bubbles Green

I sat in stunned silence, crestfallen. "That was it...?"

Our commanding general had assembled all of the company grade commanders and higher for some professional development which I definitely looked forward to. I have always loved to learn. I have always loved to read and I love academically rigorous discussion. I arrived at the assembly with no small sense of anticipation. The subject was...driver's training.

Our division had experienced a rash of vehicle accidents and the divisional leadership attributed it to an improper understanding of what a driver's training program should look like. Several hundred of us sat for a couple of hours while the division master driver explained to us what a program should look like.

"Okay, that was helpful, I guess."

Then a particular colonel stood up. At some point, a first among equals always emerges. In my circle of battalion commanders there was one officer who everyone already knew would be a brigade commander and eventually a general. He just had 'it.' Everyone knew it and it wasn't me! As an aside, he was recently selected below-the-zone for promotion. Well, the same was true of the brigade commanders. There was one that everyone just knew would be a general someday, a first among equals.

My attention focused as this particular officer stood to offer his insight. This was it. I was literally poised, pen at the ready, prepared for some deep wisdom, some enlightenment.

"I like to think of it as five 'T's," he started with.

Okay.

"You have to have the troops available to do the work."

Yes.

"You have to have the time to do the work."

That's right.

"You need the training for the soldiers to do the work."

Uh huh.

"You definitely need the tools."

Definitely.

"And most of all..."

This was it!

"...you need the task. You have to have the task. These are what has worked for me, the five 'T's of vehicle maintenance."

Again, I was crestfallen. Here was this man's chance to impart wisdom and what he had for us was...'the five T's'? I considered this a blinding flash of the obvious.

You mean I can't do the maintenance without the troops or the tools? Who would've thought?

I sat stunned, amazed at the surface-level, shallow nature of this thought. Now, this officer was unquestionably highly intelligent and extremely well educated and I'm quite positive he participated in frequent instances of deep, higher-order thought. This instance highlighted, in my mind, a rampant issue throughout much of the Army leadership that I observed: the absence of deep thought or higher order cognition.

Thoughtlessness pervades. Almost always we seemed to scratch the surface of whatever issue confronted us at the time.

Christianity is a thinking man's religion. The Bible calls the believer to be transformed by the renewing of his mind. (Romans 12:2) God calls us to question, to consider. If you truly believe what you say you believe, the blinders have finally been removed, the veil has been lifted, and your mind is no longer clouded by the presence of sin. The believer is truly a thinking man.

If you ever find yourself doubting this aspect of Christianity, I recommend you secure a copy of John Owen's *The Death of Death in the Death of Christ*—I've never actually met anyone who finished it—or spend some time in Romans chapters 9 through 11.

Thinking itself, reason and intellect, are a function of being created in the Image of God. God has given us the capacity for cognition and the Christian is to implement it, to think.

After our second suicide, one of the senior leaders in the division called my boss who was deployed at the time, "Jim, we need to stop these suicides."

My boss was speechless. How do you respond to something like that?

"Well we have, Sir. It's been two days since the last. They are definitely stopped."

As we have confined ourselves to a secular arena, as we examine the affliction of the force, we are left with nothing to do but treat symptoms, or attempt to treat symptoms. We may have no real discussion of issues and solutions. We are left merely trying to turn the red bubble green.

Our division maintained a council that I was a member of as a brigade level leader. The council sought to promote the health of the force and the community, and it consisted of several working groups There was the suicide prevention working group, the crime prevention working group, the health and welfare working group, and the sexual assault working group. During our meetings, each group lead would present the current status of their program and we'd discuss the program a bit before moving on.

At one particular meeting, the leader of the suicide prevention working group gave his spiel culminating with a discussion about the Suicide Prevention Walk they intended to sponsor. There would be a two-mile walk complete with booths and displays, all in the name of awareness and prevention of suicide. I remember thinking very skeptically that a suicide prevention walk would effectively do nothing to actually address the issue of suicide other than give leaders the ability to say, "Here, this is what we are doing about suicide. Look at our suicide prevention walk"

The sexual assault working group always infuriated me. Most recently they discussed a motorcycle ride to bring awareness to sexual assault. Again, this will likely not prevent a single sexual assault. Did they actually think that the drunken soldier alone in his barracks room with his passed out drunken female friend would ignore his erection and think to himself, "Oh yeah, the motorcycle ride...I shouldn't do this."

Please forgive my sarcasm, but this type of activity does nothing to truly address the affliction of soldiers. It does allow for the command to say, "Here, this is what we are doing," which never fails to frustrate and even infuriate me. As we are confined to the secular, we may only turn the red bubbles green.

The Army loves to codify things into charts, and over the last several years stoplight charts became de rigueur, though I don't remember when it happened. Each issue is denoted by a bubble and if it's red, that means the issue is not being handled. It's still a problem. If it's amber, then it's being worked. Green is good. The issue is addressed, no more problem. The more a commander can brief green on his charts, the better for all involved. Green bubbles become the goal aside from truly addressing the issues.

It is not as if leaders do not actually care or truly do not understand. Almost any leader will acknowledge frustration with our approach and the limitations of programs. It's just that they are confined to the secular and as such, have no other option than to focus on the bubbles, particularly because their boss wants green bubbles.

The Army's standard for readiness is 10% or less meaning that less than

10% of your forces can be non-deployable. Otherwise, commanders have explaining to do. Our division solved this problem easily by raising our standard to 8%. By meeting our own internal standard, we effectively always met the Army's standard though it changed nothing in reality other than the integrity of commanders forced to bend facts and manipulate data to obtain the requisite 8%. A public, verbal thrashing awaited any who failed to make this particular bubble green. Again, the facts on the ground never actually changed, just the color of the bubble.

As we fought the war of bubbles, the fact that soldiers actually languish in their affliction became secondary. If the bubble was the right color, all was right with the world. Never mind that soldiers continue to self-destruct independent of the color of the bubble. For them, the bubble is always red.

Mission Command

Thoughtlessness resonates in another surprising way.

In 1939, the German Army, the vaunted *Wehrmacht*, sliced through the bulk of Poland in just over a month, making short work of the defenders. Less than a year later, they would accomplish the same in France, defeating the well-prepared defenders in less than two months.

Much has been made of the combined arms maneuver capability of the *Wehrmacht*, of the concept of *Blitzkrieg* (Lightning War), and the quality of German weaponry. France actually possessed greater quantities of artillery and armor. How then had the Germans been so successful? It was the idea of *Auftragstaktik*, mission orders, that fueled the agility of the *Wehrmacht* enabling them to outmaneuver their enemies and subsequently overwhelm them.

Adopted in American mission command doctrine, the Prussians developed mission orders after defeat at the hands of Napoleon. The revolutionary concept involves the dissemination of the mission and more specifically the intent to the lowest level. Inform subordinate commanders what your intent is, what effects are desired, resource them appropriately and allow them to express initiative, and figure out *how* to accomplish the mission.

Mission orders/command relies greatly upon trust between the lower and higher echelons as much as the competency and dependability of the subordinate leaders. The initiative inherent in the concept starkly opposes previously rigorous and hierarchical implementation of orders whereby the senior commander dictates to the greatest extent possible the actions of his subordinate units.

Arab armies lose battles and wars because of a lack of agility as they cling to hierarchy. They have no bearing for subordinate leaders, for sergeants, and as such, they quickly find themselves overwhelmed by the superior mobility and agility of armies executing mission orders as fuel for combined arms maneuver. See the Six-Day War or even the Yom Kippur War for verification. It is the Arabic religion, Islam, and its subsequent devaluation of life which impedes the operational agility fueled by mission orders.

Mission orders relies upon trust and a fundamental understanding of the value of each human life and mind. I may be a general but my value to the mission is not greater than that of the squad leader. I could say that the summation of the value of the squad leaders in any conflict yields the decisive balance. Islam suppresses initiative and ingenuity, essential aspects of mission orders.

Christianity frees the mind, enabling the necessary trust in subordinates that mission orders demand. Now, the concept was developed before World War One and was firmly entrenched in German doctrine prior to Nazification and their collective descent into madness. It persisted in their doctrine which they implemented with remarkable efficiency.

Interestingly, it is Hitler's departure from mission orders that inevitably doomed the Reich. By July 1941, the *Wehrmacht* was closing on Moscow. Inexplicably, Hitler directed them to pause and deviate south, overruling his military commanders who argued for an immediate push to the Soviet capital. This 'summer pause' severely hampered the offensive as the Germans became bogged down in Kiev after encircling and capturing some 400,000 Red Army soldiers. From there to Stalingrad, the tide of the war on the eastern front turned against the Germans and they would never again regain the initiative, all as the Fuhrer violated the basic tenant that had enabled the *Wehrmacht* to be as successful as it had been.

The American military thrives on mission command, the Americanized version of mission orders. The initiative and ingenuity of subordinate leaders drives the operational agility and audacity of the combined arms team. At least, that's how it is supposed to work. I used to field phone calls from general officers like this,

"Hey Brad, General so and so, I noticed on your report that Specialist Snuffy in 1st battalion missed two physical therapy appointments but he's still on profile. What's the deal with that?"

"Sir, I'm not sure. I'll have to get back to you."

As the Army has become increasingly paranoid about readiness and answering to its civilian masters about the affliction of soldiers, leaders have increasingly abandoned the mission command that our very doctrine centers around. This abandonment has its roots in trust, or lack thereof. Leaders, fearful of failure and reprisal, simply do not trust subordinates at some level.

Now, obvious exceptions exist. My last boss was an intense mission command leader. I would go weeks without speaking to him and then start to feel guilty and give him a call to let him know we were still doing stuff, still executing his intent.

"No problem, Brad. I've been keeping track."

It has seemingly not occurred to some of the senior leaders that accepting a bit of risk on behalf of junior leaders actually bolsters the organization as it strengthens trust and increases the competence of those same junior leaders. Mission command functions best in a climate of trust yet micromanagement permeates the Army, at least the part of the Army that I have observed.

I blame the darkening of minds and the abandonment of true knowledge for secular solutions that actually provide very little in the way of value. Godlessness foments mistrust at every level, anathema to mission command.

A Darkened Mind

Often us Generation Xers gaze upon the Millennials that populate the ranks and shake our heads in astonishment. Millennial thought sometimes seems so alien, and so you'll see hear many of my peers bemoaning the current state of things and wistfully exalting the 'old days' when men were truly men. Millennials, just like us, are a product of a number of factors. Various social factors collaborated with the degradation of the Church in yielding what we now decry. I ask, who allowed the Church to fade from relevance? Who first walked from the Church? In many cases, it is the Millennial's parents who first walked way...us.

We find ourselves too busy denouncing conditions and implementing secular methodologies than to worry about reclaiming Christ's position in our culture. Perhaps it's because many of us don't know Him ourselves. And as sin clouds the minds of our leaders, it likewise clouds the minds of our soldiers. Sin just looks a bit different in the minds and actions of the unchurched vice that of the de-churched.

A subtle frailty permeates the ranks, a fragility of mind that I closely associate with the permeation of sin. As I prepared to come to the division, a friend of mine who just completed a tour in the same division cautioned me, "Be careful how you talk to your young officers."

He warned me that the slightest harsh word or even glance or facial expression would shatter the psyche of these young officers. Our community was notorious for bitter and harsh debriefs of operations, and if you sucked you were told that you sucked. And if you couldn't handle that, then you had to seek life elsewhere. Not the case in the division. Young soldiers in particular could not handle the stressors of daily life, much less the stress of combat.

The Army devotes a significant quantity of resources to bolster resilience. We ever seek ways to improve the soldiers' ability to handle and respond to stress, to inoculate them against stress. Despite this, many of them cannot cope. The slightest rough spot, even if it is of their own doing, drives them to unravel, come apart at the seams. Many end up in the local mental health facility for just not being able to cope. Stress them out and they'll issue a suicidal ideation necessitating a two-week stay at the facility in a stress-free environment where nothing is demanded of them.

My mother faithfully maintained a few of my high school memorabilia including a few VHS tapes of football games (I'm aware that I'm dating myself). During one visit home, my girls and I perused one of the games. I had a coach that had a particular affinity for me named Coach Ferguson or 'Ferg'. At one point during the game, though they were filming from the press box, you could very clearly hear Ferg screaming at the top of his lungs, "SMITH! SMITH! WHAT ARE YOU DOING! SMITH!"

I remember very distinctly the sharp retort of Ferg's whistle as it made contact with my helmet as he sought to reinforce a certain point on occasion.

The pinnacle of soldierly pursuit has become the permanent status of victimhood. Can I obtain a diagnosis, particularly of PTSD, that will allow me to permanently claim victimhood and all that accompanies this claim—sympathy, a paycheck for life. It is not lost on the soldiers that a PTSD diagnosis will inevitably yield a 70% disability claim and though PTSD is a real thing, I recently learned of a newer diagnosis, that of garrison PTSD. Fewer and fewer soldiers are submitted to the rigors of combat, but the diagnosis persists and even pervades.

The pursuit of victim status, a frailty of the mind, and a shamelessness at defrauding the government weakens our Army and dishonors those who actually do struggle. The Integrated Disability Evaluation System (IDES) evaluates soldiers on whether or not they can continue to serve based upon their medical condition. A deluge of soldiers seeking the correct labeling floods the system, overwhelms it.

Many soldiers openly pursue a medical board as in, "I think I'm going to shoot for a medical board." They consort and collaborate in the barracks. If you say the right things in the right way to the right people, perhaps you too could obtain a medical separation and the coveted paycheck for life. Many soldiers enter service believing themselves already entitled.

It is with utter confidence that I know that I could contact the appropriate behavioral health specialist and after a few visits, obtain a PTSD diagnosis and perhaps a nice prescription for some psychotropic meds. I'd automatically be enrolled in IDES. Next, I'd follow that up with some repeated consults for unspecified back pain and again, I'd probably be issued a nice prescription for narcotic pain medication. I'd appeal my initial VA ratings in the IDES system as unjust and perhaps add some diagnoses during the process, necessitating a restart so that I could milk the system that much longer. The longer I can remain on active duty the better since I draw full pay but don't actually have to do anything because I have a restrictive profile and if my leaders try to force me to do anything, I'll have a suicidal episode and go play ping-pong and basketball in the local mental health facility for two weeks.

This sounds like a joke or an isolated incident, but it's not. This happens often and it's sin, nothing more. We allow it as a nation and as an Army. We enable soldiers and in effect, enable sin. The system enables sin. The ramifications resonate in the foundational weakness of the force. I pray we don't get drawn into a cutthroat dog-fight anytime soon.

These are not just the young soldiers anymore. These are now the squad leaders, the sergeants, the section sergeants. These soldiers now occupy the most critical position in the Army, the first line leader. A good unit led by good squad leaders, led by a good commander, can accomplish anything. A unit with good squad leaders can still function even if encumbered by a bad commander. However, a unit with uncommitted, frail squad leaders will crumble were Patton himself in charge.

And it's our fault, the leaders of this nation and the leaders of this Army. We've allowed Satan into the ranks by removing Christ from the ranks, and we've combated this condition with secular methodologies that often exacerbate whatever condition we've sought to address.

Again, I fall short in answering, "What do we do about it?" Short of once again offering, "Repent!"

Again, the question morphs. "What do we do about it?" becomes,

"How do we overcome this dearth of godless behavior with secular solutions," or

"How do we stop soldiers from acting like godless heathens apart from the only proven way, the Gospel of Jesus Christ?"

I profess to not knowing and again and can only pray for repentance, revival in the nation, and in the ranks.

The Parting Glass

A good Army leader never identifies a problem to his superior officer without identifying a corresponding solution. This is Army 101, taught from the earliest days. As I realize the last several chapters have introduced a litany of problems with very little in the way of solution, I don't want to come off as entirely negative.

I maintain a deep and professed love for the Army and for soldiers. It is the pain in my heart at seeing the affliction of soldiers and the erosion of the proud and strong foundations of my Army that have compelled me to write. What is the soldier to do? What is the leader to do?

I'll return to my previous exhortations. Are you of Christ? If so, are you a silent Christian? Do you quietly sit by as injustice and godlessness pervades? Will you take a stand? Too many men sit on the sideline as things that they know are wrong wreak havoc upon the masses lest they get involved and endanger their own standing. Realize that a vast struggle pervades. I exhort you: allow Christ to rule through your actions. Serve as an agent of Christ. Lead as an agent of Christ. Realize that He has entrusted to you the ministry of reconciliation and seek to reflect Christ in all that you do, particularly in your Army service.

Will you, believer, take a stand for Christ no matter the cost? Our nation desperately needs strong men to stand and reclaim this proud country on behalf of the Lord, to resolutely declare, "Not on my watch," to stand against the iniquity of the world. Our soldiers are desperate and hungry for that which they could never fathom. They are desperate for godly leaders and for godly men to show them the way. Would you be that man?

If you are not of Christ, then you have an altogether more pressing concern.

Final Fires

Sunday, July 10

I feel I must write quickly before fatigue overtakes me. It's nearly 9:00 p.m. and I find myself but where else, flying commercial out of Iraq, something I was sure I would never do. I must say that it is slightly surreal to transition immediately from the conduct of war to walking into an airport and boarding a commercial jet.

As much as I like to fly military air, I now remember another reason I don't enjoy flying commercial. The child sitting in front of me is in desperate need of a rather good dose of discipline and his mother went full recline on me without a preparatory move and literally banged me in the forehead with the back of the seat. As this woman is practically laying in my lap, I'll necessarily use my T-Rex arms to finish this.

My war is over I believe. It might be a bit premature, but I'm calling it. I've no fight left in me or maybe I do, but nevertheless, it's over. I still have a year or so left on duty so it is possible they might try to squeeze more blood from this turnip, but it's unlikely.

This last deployment was slow but fruitful in a number of ways. As always, I am very ready to get home to see my beautiful wife and my people. I am determined that this time, things will be different.

I've noticed with other friends that homecoming is both a time of immense joy and trepidation or stress, particularly for repeat deployers. Me and Ami's reunions follow a certain routine. The first day or two are great: smiles, hugs, and tears of joy. Yet, within a short period of time, I start to notice things not to my liking. Maybe she rearranged a room or maybe the house is messy, something. I'll pretend that everything is fine.

I'm a lousy pretender and at some point I actually want Ami to know I'm dissatisfied. Eventually, I'll cave and confess to being unhappy with something. At this point, she cries and I feel bad for making her cry but actually better that I've gotten to air my dissatisfaction. Aren't I just a great guy? I think it goes similarly for others. Imagine the bitter contrast between expectation and reality.

This is the big moment, the reunion, what you've both been looking forward to for so long and you want it to be perfect. She wants her knight in shining armor to ride in out of the sunset, sweep her off her feet once more, and carry her away from all that she has been dealing with. You likely want to come home to things as they were, an orderly home and willing wife ready to simultaneously jump in the sack and surrender her role in the home. The reality is very different.

Though I've never been in her shoes, I can imagine her perspective. She's stayed behind, taken care of the children and home, everything and all that implies, all while not knowing what I'm doing. Am I safe? Will I come home again? Reports of death and fighting in the news keep her awake at night. Two personal incidents highlight this issue.

Just last week, my wife started the van and it began to smoke. Ami turned it off, opened the hood, and started it once more noting that the smoke seemed to be coming from the alternator. She had it towed to our mechanic's shop at the front of our neighborhood, had them remove the alternator, took it to the AutoZone since it was still under warranty as I had just replaced it a few months ago. The AutoZone replaced it, she took it back to our mechanic who installed it. Problem solved, $100, less than two hours. She does things like that routinely, even while I'm home, but daily while I'm deployed. Oh yeah, she does that with five little monkeys hanging from her legs.

The other incident occurred when Worrell and Weeks died. When a death occurs down range, they shut off all communications until they can notify the families. They always know, from the news or wherever, the families always know that someone has died and they brace for the news, praying it's not their man.

Everyone knew there had been a gunfight and that two had died. Ami was at the YMCA running on a treadmill, crying, and praying that it wasn't me. I can't imagine the agony of not knowing whether the one you loved was safe. Well, the Lord sent a messenger. One of my guys happened to be at the YMCA that day and saw Ami and quietly let her know that I was okay.

These are the things our wives deal with and they just want us to come home and love them and love being home in spite of things not being exactly how we'd like them. Why would we care as long as we were home with our families? No wonder the appearance of unhappiness or dissatisfaction is so hurtful.

On the other hand, the soldier has a similar challenge. While deployed, a soldier works, sleeps, eats, and maybe works out. Over there, life is incredibly simple and organized. For the last two months, my entire sphere of existence outside of work was the couple square feet next to my bunk. No mess. No clutter. Life is simple, almost easy. Everyone follows orders for the most part and does what they are supposed to do. Outside of work, I am responsible for only me. Each night, I read a bit and then get much more sleep than I do at home. No babies crying. No diapers to change. No kids fighting. No drama, other than the obvious stresses of combat itself.

Usually, a few days before I come home, Ami will send me an email detailing our upcoming events. There is no grace period, no easing back into the busyness. You step off the plane and it's full speed back into the sprint of life.

This trip, I'll get home on Monday evening and watch the boys Monday night while Ami goes to a meeting. Our in-laws will already be in town for my daughter's wedding on Saturday. They are staying at our home. I'll maybe take Tuesday off and will be back at work on Wednesday. I took some leave before I left so I'll get no time off. Right back to work. Somewhere I'll have to catch my breath as the crazy busy pace of our lives threatens to overwhelm me.

This time though, I'm determined that things will be different. I am determined not to make Ami cry. I'm determined to show the grace that she so deeply needs and deserves and to focus on the things that matter. I pray that on this last return, I'll finally get things right.

I'm tired. I can truly say that I'm tired of fighting. It is with much relish that I lay down this sword. The fire in my chest to close with and destroy the enemy has become but an ember. I know that another fight awaits.

For that, I'll necessarily turn to the Fountain of living water that is the Lord Jesus. I'm confident His grace will be more than sufficient.

—Soli Deo Gloria—

Bibliography

American Foundation for Suicide Prevention. https://afsp.org/about-suicide/suicide-statistics/
    (accessed October 14, 2016)

American Psychological Association. http://www.apa.org/topics/ptsd/index.aspx
    (accessed May 30, 2016).

Barajas, Joshua. "How the Nazi's defense of 'just following orders'
    plays out in the mind." *PBS NewsHour*. http://www.pbs.org/newshour/rundown/how-the-nazis-defense-of-just-following-orders-plays-out-in-the-mind/ (accessed May 26, 2016).

Bare, Stacy. "The Truth About 22 Veteran Suicides a Day." *Task and Purpose*.
    http://taskandpurpose.com/truth-22-veteran-suicides-day/ (accessed October 14, 2016).

Castillo, Michelle. "Almost half of first babies in U.S. Born to unwed mothers." *CBS*
    News. http://www.cbsnews.com/news/almost-half-of-first-babies-in-us-born-to-unwed-mothers/ (accessed October 14, 2016).

Chaggaris, Steve. "'I couldn't be prouder' of Jason Collins, Obama says." *CBS*
    News. http://www.cbsnews.com/news/i-couldnt-be-prouder-of-jason-collins-obama-says/ (accessed May 24, 2016).

Childress, Sarah. "Why Soldiers Keep Losing to Suicide." *Frontline*.
    http://www.pbs.org/wgbh/frontline/article/why-soldiers-keep-losing-to-suicide/ (accessed June 10, 2016).

Crane, Stephen. *The Red Badge of Courage*. http://www.emcp.com/previews/AccessEditions/ACCESS%20EDITIONS/The%20Red%20Badge%20of%20Courage.pdf, 42 (accessed May 30, 2016).

Cole, Matthew. "Al Qaeda Promises U.S. Death By A 'Thousand Cuts'." *ABC News*. http://abcnews.go.com/Blotter/al-qaeda-promises-us-death-thousand-cuts/story?id=12204726 (accessed March 29, 2016).

Cummins, Joseph. *The World's Bloodiest History*. Beverly: Fair Winds Press, 2009.

Dao, James. "Drone Pilots Are Found to Get Stress Disorders Much as Those in Combat Do." *The New York Times*, http://www.nytimes.com/2013/02/23/us/drone-pilots-found-to-get-stress-disorders-much-as-those-in-combat-do.html? r=0 (accessed June 11, 2016).

Dao, James and Andrew W. Lehren. "Baffling Rise in Suicides Plagues the U.S. Military." *The New York Times*. http://www.nytimes.com/2013/05/16/us/baffling-rise-in-suicides-plagues-us-military.html (accessed July 12, 2016).

David, A.R. *The Pyramid Builders of Ancient Egypt: A modern investigation of Pharaoh's workforce*. London: Guild Publishing, 1986.

Deffinbaugh, Robert L. "Human Responsibility and Salvation." *Bible.org*. https://bible.org/seriespage/12-human-responsibility-and-salvation-romans-930-1021 (accessed March 17, 201).

DSM-V. http://www.dsm5.org/Documents/PTSD%20Fact%20Sheet.pdf (accessed May 30, 2016).

Eck, Werner. *The Age of Augustus*. Oxford: Blackwell Publishing, 2003.

Flavius Josephus. *Antiquities of the Jews, II-IX*-I. 2016, http://www.biblestudytools.com/history/flavius-josephus/antiquities-jews/book-2/chapter-9.html (accessed April 4, 2016).

Frances, Allen. "Dying Well Means Dying at Home." Psychiatric Times. http://www.psychiatrictimes.com/couch-crisis/dying-well-means-dying-home (accessed May 29, 2016).

Friedman, Matthew J. "History of PTSD in Veterans: Civil War to DSM-5." http://www.ptsd.va.gov/public/PTSD-overview/basics/history-of-ptsd-vets.asp (accessed May 30, 2016).

Gao Xingzu, Wu Shimin, Hu Yungong, Cha Ruizhen. *Japanese Imperialism and the*
Massacre in Nanjing, Chapter X, Widespread Incidents of Rape. http://museums.cnd.org/njmassacre/njm-tran/njm-ch10.htm on 11 (accessed March 11, 2016).

"Guilt". *Encyclopedia of Psychology.* 2nd ed. Ed. Bonnie R. Strickland. Gale Group,
Inc., 2001. eNotes.com. 2006. 31 December 2007. http://www.enotes.com/ homework-help/psychology-what-guilt-what-stages-guilt-466309 (accessed May 18, 2016).

Grimal, Nicolas. *A History of Ancient Egypt.* Wiley-Blackwell, 1994.

Grossman, Dave. *On Killing.* New York: Back Bay Books, 2009.

Grudem, Wayne. *Systematic Theology: An Introduction to Biblical Doctrine.* Grand
Rapids: Zondervan, 1994.

Guttman, Stephanie. *The Kinder, Gentler Military.* New York: Scribner, 2000.

"Heart." *Baker's Evangelical Dictionary of Biblical Theology.*
http://www.biblestudytools.com/dictionary/heart/ (accessed April 18, 2016).

"Heart." *Strong's Concordance*, Biblehub.com. http://biblehub.com/greek/2588.htm
(accessed May 13, 2016).

Helsel, Phil. "Caitlyn Jenner Receives ESPY Arthur Ashe Award for Courage." *NBC News*. http://www.nbcnews.com/news/us-news/caitlyn-jenner-receives-espy-award-courage-n392911 (accessed May 24, 2016).

Hennessy, John J. "War Watchers at Bull run During America's Civil War." *HistoryNet*. http://www.historynet.com/war-watchers-at-bull-run-during-americas-civil-war.htm (accessed June 3, 2016).

Hitler, Adolf. "Race and People." *Mein Kampf*. https://www.stormfront.org/books/mein_kampf/mkv1ch11.html (accessed April 14, 2016).

Hochschild, Adam. *To End All Wars - a Story of Loyalty and Rebellion, 1914-1918*. Boston, New York: Mariner Books, Houghton, Mifflin Harcourt, 2012, accessed May 30, 2016, http://library.umd.umich.edu/newbooks/2012/may.php.

Jones, Graham. "Srebrenica: A Triumph of Evil." *CNN*. http://www.cnn.com/2006/WORLD/europe/02/22/warcrimes.srebrenica/ (accessed May 13, 2016).

Judgment International Military Tribunal for the Far East. http://www.ibiblio.org/hyperwar/PTO/IMTFE/IMTFE-8.html (accessed March 11, 2016).

Kaiyuan Zhang. *Eyewitness to Massacre: American Missionaries Bear Witness to Japanese Atrocities in Nanjing*. M. E. Sharpe, 2001.

Keegan, John. *The American Civil War*. New York: Knopf, 2009.

Klooster, F.H., "Sovereignty of God." in *Evangelical Dictionary of Theology*. 2nd Edition. Edited by Walter A. Ewell. Grand Rapids: Baker Academic, 2001.

"Leadership According to Eisenhower." http://leadership.w9z.org/leadership-according-to-eisenhower/ (accessed May 13, 2016).

Mayo Clinic. http://www.mayoclinic.org/diseases-conditions/post-traumatic-stress-
   disorder/basics/definition/con-20022540 (accessed May 30, 2016).

Martin, Courtney E. "Zen and the Art of Dying Well." *The New York Times*.
   http://opinionator.blogs.nytimes.com/2015/08/14/zen-and-the-art-of-dying-well/ (accessed May 29, 2016).

McNab, Chris. *Hitler's Master Plan*. Amber Books, 2011.

"Michael Sam earns national praise, support." St. Louis Post-Dispatch.
   http://news.live.stltoday.com/Article/360104-Michael-Sam-earns-national-praise-support (accessed May 25, 2016).

Michaels, Jim. "Marine study finds all-male infantry units outperformed teams with
   Women." *USA Today*. http://www.usatoday.com/story/news/nation/2015/09/10/marine-study-finds-all-male-infantry-units-outperformed-teams-women/71971416/ (accessed October 14, 2016).

National Center for PTSD. http://www.ptsd.va.gov/professional/PTSD-overview/index.asp (accessed May 30, 2016).

National Institute of Mental Health. http://www.nimh.nih.gov/health/topics/post-traumatic-stress-disorder-ptsd/index.shtml (accessed May 30, 2016).

"Neville Chamberlain on Appeasement (1939)." *The History Guide*.
   http://www.historyguide.org/europe/munich.html (accessed March 24, 2016).

Pearson, Michael. "The Petraeus affair: A lot more than sex," CNN.com.
   http://www.cnn.com/2012/11/12/us/petraeus-cia-resignation/ (accessed September 25, 2017).

Piper, John. "Total Depravity." https://www.monergism.com/thethreshold/articles/piper/depravity.html (accessed on December 7, 2016).

Schafer, Peter. *The History of the Jews in Antiquity*. Routledge: New York, 1995.

Shakespeare, William. *Henry V*. http://quotationsbook.com/quote/3024/ (accessed June 13, 2016).

Shudo, Higashinakano, Kobayashi Susumu and Fukunaga Shainjiro. "*Analyzing the*
> *'Photographic Evidence' of the Nanking Massacre (originally published as Nankin Jiken: "Shokoshashin" wo Kenshosuru)."* Tokyo, Japan: Soshisha, 2005 http://www.sdh-fact.com/CL02_1/26_S4.pdf (accessed March 11, 2016).

Sun-Tzu, *The Art of War*. New York: Barnes and Noble Books, 1994.

Starr, Penny. "*New Miss USA and Army Vet: 'We are Just as Tough as Men'*." *CNS*
> *News*. http://cnsnews.com/blog/penny-starr/new-miss-usa-and-army-vet-we-are-just-tough-men (accessed October 14, 2016).

"The Heroes of My Lai." http://law2.umkc.edu/faculty/projects/
> ftrials/mylai/myl_hero.html (accesssed October 14, 2016).

"The Rise of Adolf Hitler." The History Place. http://www.historyplace.com/
> worldwar2/riseofhitler/mother.htm (accessed April 18, 2016).

Veterans and PTSD. "Veterans statistics: PTSD, Depression,TBI, Suicide."
> http://www.veteransandptsd.com/PTSD-statistics.html (accessed May 28, 2016).

Von Clausewitz, Carl. *On War*. Edited by Anatol Rapoport. London: Penguin Books, 1968.

Walter, V.L. "Arius, Arianism." in *Evangelical Dictionary of Theology*. 2nd Edition.
> Edited by Walter A. Elwell. Grand Rapids: Baker Academic, 2001.

"Westminster Confession of Faith." Center for Reformed Theology and Apologetics.
> http://www.reformed.org/documents/wcf_with_proofs/ (accessed March 18, 2016).

Wilson, Mar. "Number of US army officers fired for misconduct tripled in 3 yrs—
> report." Reuters. https://www.rt.com/usa/army-misconduct-soldiers-rise-280/ (accessed September 25, 2017).

Wong, Leonard and Stephen J. Gerras, "Lying to Ourselves: Dishonesty in the Army
> Profession." *Strategic Studies Institute.* http://www.strategicstudiesinstitute.
> army.mil/ pdffiles/PUB1250.pdf (accessed May 21, 2016).

Woods, John E. *The Good Man of Nanking, the Diaries of John Rabe.* 1998.

Yang, Celia. "The Memorial Hall for the Victims of the Nanjing Massacre: Rhetoric
> in the Face of Tragedy" (2006). http://bootheprize.stanford.edu/0506/PWR-Yang.pdf (accessed March 11, 2016).

Ye Hee Lee, Michelle. "The missing context behind the widely cited statistic that
> there are 22 veteran suicides a day." *The Washingon Post.* https://www.washingtonpost.com/news/fact-checker/wp/2015/02/04/the-missing-context-behind-a-widely-cited-statistic-that-there-are-22-veteran-suicides-a-day/ (accessed May 24, 2016).

BRAVE RIFLES
The Theology of War

Deep Dive Companion Bible Study

By
Bradford Smith

About this Study Guide
The Deep Dive Companion

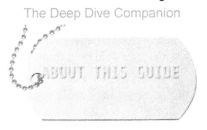

For though by this time you ought to be teachers, you need someone to teach you again the basic principles of the oracles of God. You need milk, not solid food, for everyone who lives on milk is unskilled in the word of righteousness, since he is a child. But solid food is for the mature, for those who have their powers of discernment trained by constant practice to distinguish good from evil.
Hebrews 5:12-14

Welcome to the Deep Dive Bible Study series. This work is designed as a companion for *Brave Rifles: The Theology of War.*

At some point, you've wanted something more, something deeper. Foster says,

Superficiality is the curse of this age.

Christianity is a thinking man's religion and each time I dive into the text, I become aware of how shallow my previous excursions have been. As God's revelation of Himself to men, the Bible possesses limitless capacity to transform a man by the renewal of his mind. (Romans 12:2)

Ami and I at some point became slightly disgruntled with the nature of many contemporary Bible studies and their inability or unwillingness to handle the toughest and deepest questions. These shortcomings come to particular light when handling difficult questions like those concerning evil and suffering in light of the sovereignty of God.

Our home group was studying Matthew 6 whereby Jesus assures the audience from His Sermon on the Mount that He would take care of them. Three times He exhorts them, "Do not be anxious." (v. 25, 31, 34) He would provide everything for them. Doesn't He care for the birds of the air or the lilies of the field? How much more would He provide for His people.

In the following verses, He urges His people to ask and that it would be given to them. Whoever asks receives. What kind of father would give his son a stone when he asks for bread? What kind of father would give his son a serpent when he asks for a fish? "How much more will your heavenly Father who is in heaven give good things to those who ask him!" (Matthew 7:11b)

From these texts, the home group leader surmised that God in fact is good and gives us all that we need. Ami raised her hand and asked a question,

"What about Christians who starve to death, or are raped, or have horrible things happen to them? How is God giving them all that they need?"

Now, she wasn't asking to be snarky or confrontational. She wanted to know the answer to this because it was a very relevant question in her life at the time. She was dealing with some personal, family issues. One of our sons had been diagnosed with epilepsy and seemed to get a new diagnosis monthly as his health deteriorated. Our oldest son had disappeared back into the Memphis streets that we had adopted him from, seemingly consumed by his former afflictions.

Additionally, she had come from a very difficult background and childhood including all manner of abuse and abandonment. I can only imagine the depth of her residual trauma. With all of these things, hearing about God's gracious abundance presented a seeming contradiction to her concerning God.

They added a dose of reality to the teaching regarding the provision of God. How does God provide when His people so often suffer, frequently at the hand of others? In fairness, the home group leader was not exactly prepared to answer these difficult questions and I've found that many assigned the important role of teaching in the church, likewise are challenged to answer these questions.

It becomes apparent at some point, that Christians, including leaders, are content to consume milk and not tear into the meat of Scripture. As such, you've heard some of these epithets, I'm sure,

> *Just have faith.*
> *Trust Jesus.*
> *Let go and let God.*

Now, I'll not deny the truths of these exhortations, but there is so much more beneath the surface, especially when you start considering Scripture in light of the difficult situations of life and in this case, that of suffering.

If we never make that leap, then our pursuits remain cold and academic. Only by reconciling Scriptural truth with life, an often difficult prospect, do we find the transformation that most of us so desperately need and desire.

My prayer is that the Deep Dive Bible Study Series starting with this work might be a blessing to you. I pray that you would not be content with a shallow, surface-level pursuit of God, that this study might just fan the flames of your hunger for His word, to Him be the glory.

Study Guide Methodology

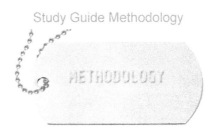

The *Deep Dive Companion* Bible Study to *Brave Rifles* serves as a guide to focus your studies and guide personal meditation on the material. It can be used for personal study or a small group.

Ideally, the student will read each chapter in Brave Rifles and then answer the corresponding questions pertaining to that chapter in this *Deep Dive Companion* Bible Study. Most of the Bible references are quoted in their entirety in the accompanying chapter in *Brave Rifles*.

Each section begins with a question labeled '**Benchmark**.' These are questions that have a particularly relevant focus for the student. The **Benchmark** questions are intended to be reviewed and considered as the student moves into subsequent sections.

The very last section is a review of the **Benchmark** questions.

Introduction

Study Guide for
Brave Rifles: The Theology of War

INTRODUCTION

The society that separates its scholars from its warriors will have its thinking done by cowards and its fighting done by fools...Thucydides 460-400BC

Preparatory Fires

Brave Rifle,

I don't know what prompted you to pick up this study or the book. I pray it is because you are searching for something more, something real and tangible. Maybe you've fought. Maybe you've witnessed the awesome tragedy of warfare between men first-hand. Maybe you've seen what modern munitions do to a man or employed them yourself.

Perhaps you've engaged in questionable activity in a time of war or you have your own questions. I assure you, answers exist. They are not easy answers. You must open your heart to them.

Do you bear scars from war from a singular episode, or possibly years of persistent combat? Maybe you have scars upon scars and maybe they have no visible manifestation. They remain buried deep within your soul. Or, perhaps they've overflown into your life and manifested themselves in a number of dysfunctional ways. Do you suffer from your scars?

There is a Healer, and it is the pursuit of truth that will ultimately take you before Him. You may deal with the scarring in other ways, but I ask, "Do you want to get well?" I pray that you do. God bless you in this pursuit.

Brad

Study Guide for Chapter 1

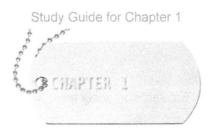

Approaching Jericho

*Now these are the nations that the LORD left, to test Israel by them, that is, all in Israel who had not experienced all the wars in Canaan. It was only in order that the generations of the people of Israel might know war, to teach war to those who had not known it before. These are the nations: the five lords of the Philistines and all the Canaanites and the Sidonians and the Hivites who lived on Mount Lebanon, from Mount Baal-hermon as far as Lebo-hamath. They were for the testing of Israel, to know whether Israel would obey the commandments of the LORD, which he Commanded their fathers by the hand of Moses. So the people of Israel lived among the Canaanites, the Hittites, the Amorites, the Perizzites, the Hivites, and the Jebusites. And their daughters they took to themselves for wives, and their own daughters they gave to their sons, and they served their gods.
Judges 3:1-6*

Benchmark: What do desire to get out of this study? What are your objectives? Why? We'll examine these at a later date?

1. What is your experience with war? Have you personally deployed? Participated in direct combat? Have a family member who has experienced combat?

2. List some impressions you have of Jesus? What are some popular ideas concerning Jesus?

3. Where do most people get their ideas and information concerning Jesus? Why these sources?

4. What purposes does war serve, in either antiquity or present times? What have you observed?

5. What are popular culture views concerning war? How do they compare to the reality as you know it?

6. Read Matthew 5:46. Define common grace as you understand it.

7. Read Judges 3:1-6.

    a. What are the two purposes given that God left the nations?

    b. Why did God need to leave them, for what purposes?

8. This work will explore the applicability of these two concepts to modern times. What are your initial thoughts concerning the two reasons as they pertain to us today?

Study Guide for Chapter 2

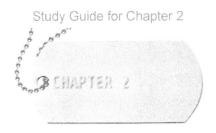

Context of the Examination

Benchmark: How would you consider your faith, currently? In what do you place your faith? Do you have any faith? If you are a Christian, how actively do you practice your faith?

1. Read 1 Thessalonians 5:16-18, Galatians 5:24, and Matthew 16:24. What do these verses tell us about the Christian faith?

2. What does Jesus mean in Luke 9:60-62?

3. Read Luke 14:26. What does Jesus mean? Does He truly desire one to hate his father and mother or wife and children? Why does He say this?

4. Read Colossians 3:23. What are the implications of this verse? What are the implications to your daily life?

5. The majority of Americans profess Christianity yet statistics show steeply declining measures of Christian practices such as church attendance and daily Bible reading. Why is this? Do you think most Christian apply the previous verses? Why or why not?

6. Read Matthew 5:44, Matthew 5:39, Luke 6:27, and Luke 6:36. What does Jesus mean in these verses? Are these easy things to do? What stands in the way of us behaving in this way?

7. How do you view the Bible? What are your thoughts on its reliability? Where do you get your ideas concerning the Bible? Have you read it? How often do you read it?

8. From Chapter 2, define the historical-grammatical method of biblical interpretation. Why is interpretive method important?

7. What are the possible conflicts between Jesus' teachings and the conduct of warfare? Has this ever affected you personally? Why or why not?

8. What questions do you have concerning war and its conduct?

9. Can you identify any initial difficulties in reconciling the Bible with the conduct of war?

Study Guide for Chapter 3

First Things, Hardest Things (part 1)

Benchmark: What are your thoughts about or impressions of God? His character? His attributes? His work? From where do you get your ideas concerning God?

1. Read Romans 11:33 and Isaiah 55:8-9. What do these two verses tell us about God?

2. Read Proverbs 1:7. From where does all knowledge begin? Why "fear"? Do you "fear" God in the sense of this verse?

3. From Chapter 3, why is it necessary and useful to conduct the study of war from a Christian perspective?

4. Rewrite the following phrase in your own words, "Orthodoxy always precipitates orthopraxy."

5. Do you attend church? What is your church's statement of faith concerning the Bible and its inerrancy?

6. In your opinion, can we trust the Bible? Why or why not? Review what Chapter Three says concerning the matter.

7. Read Colossians 1:15 and Hebrews 1:3. What are some descriptions of Jesus given? From Chapter Three, define the hypostatic union.

8. Why do many people reject the orthodox understanding of Jesus in favor of something different? What are your views concerning the hypostatic union? Why?

9. What are your initial thoughts concerning the effectiveness of an army of believing Christians? More or less effective? Why?

Study Guide for Chapter 3

First Things, Hardest Things (part 2)

Benchmark: Review your benchmark answers from last week. Can you identify anything in your life that contributed to your ideas concerning God?

1. From Chapter 3, define the sovereignty of God.

2. Spend a few minutes listing *things* and *actions*, from the largest to the smallest, the most complex to the simplest. Do you believe that God's sovereignty includes these things? What are the implication of this?

3. Read aloud 2 Chronicles 20:6, Psalm 115:2, Psalm 135:6, Isaiah 45:7, Romans 11:36, and Colossians 1:17. As you read, meditate upon the truths of these verses.

4. From Chapter 3, define providence.

5. From Chapter 3, discuss free will. What does it mean? What do most people mean when they say free will? Does ungoverned (libertarian) free will exist? Why or why not?

6. From Chapter 3, define concurrence.

7. How does Joseph respond to his brothers' plea for forgiveness?

8. Read Isaiah 10:6,7. Isaiah is writing about the Assyrians, a cruel nation threatening the northern kingdom of Israel. From the verse, what is God's command to Assyria? What does it mean when he says, "But he does not so intend?"

9. Read Romans 8:28. What are the implications of this verse?

10. What was the wickedest act in history? Who directed and executed this act?

11. Read Deuteronomy 29:29. What are the implications of this verse?

Study Guide for

Section 1: War and the Body

Brave Rifles: The Theology of War

Study Guide for Chapter 4

The Source of War

Benchmark: What is the origin of war?

1. What are your thoughts concerning the following statements? Do you concur? Why or why not?

a. "No country has ever profited from protracted warfare."

—Sun-Tzu

b. "War, therefore, is an act of violence intended to compel our opponent to fulfill our will."

—Clausewitz

c. "Thus what (motivates men) to slay the enemy is anger; what (stimulates them) to seize profits from the enemy is material goods."

—Sun-Tzu

d. "In short, even the most civilized nations may burn with passionate hatred of each other."

—Clausewitz

2. Read Genesis 1:26 and 2:7. How is man distinct from the rest of creation?

3. Genesis Chapter 3 records the Fall of man. (*Extra study: read Genesis 3*) Read Romans 5:12. What were the results of Adam's transgression?

4. Read Romans 8:20-22. According to Paul, what happened to creation? Why does this matter in our study of war?

5. From Chapter 4, what is the first recorded *act* as a function of the Fall of man? What were Cain's motives?

6. Read James 4:1,2. What is the source of quarrels and fights according to James?

7. Read Genesis 6:5-11. What four things condemned the earth to God's judgment?

8. In your own words, based upon what you've read thus far, what is the origin of war?

Study Guide for Chapter 5

Road to War

Benchmark: What is your understanding of the overall theme of the Bible? Is there one?

1. Read 1 Corinthians 1:18. What does this verse mean? Is there any benefit for a non-believer in reading the Bible?

2. Read Hebrews 4:12. How does the author of Hebrews describe the Bible? What are the implications of this?

3. From Chapter 5, what is God's initial command to Abraham? What are the three aspects of God's covenant with Abraham?

4. Who occupied the land promised to Abraham's descendants? What kind of people were they? Of what were they guilty? What did God do?

5. How did God's people end up in Egypt?

6. What was the consistent report of the Israeli spies concerning Canaan? Who dissented? What did they say? What was the outcome of this incident?

7. Read Deuteronomy 7:1-2. As this seems excessively harsh, how do you reconcile this command with a just, merciful, and loving God?

8. Who led the invasion? Why? Where was Moses?

9. How did the invasion go? What happened? Why? What was the outcome?

10. How long did it take for the people to forsake God? What was God's response? Why did the people not complete the conquest?

Study Guide for Chapter 6

War and Sovereignty

Benchmark: Describe a time you've been afraid in battle. If not in battle, a time that you've been fearful. What generated the fear? What did you do? How did you deal with the fear?

1. Read Romans 8:28. Which things work together? Does this include war? What are the implications of this?

2. From the previous sections, define in your own words sovereignty, providence, and concurrence.

3. Read Judges 3:1. From this verse, who left the nations in Canaan? For what purpose?

4. How did Israel hold up in this testing from God? What happened?

5. What are God's repeated commands to Israel prior to the invasion? Why?

6. How do the men of Israel prepare for battle? What is the modern-day corollary to this? How does or should the warrior today prepare for battle?

7. If you've been to combat, recall the longest day you've had, the toughest fight? What made it so long and difficult? What did you do? How did you respond? What about those around you?

8. What are the two lessons concerning *testing* from Chapter 6?

9. What were the consequences for Israel's failure?

10. How does Chapter 6 define war? What are your thoughts on the matter? How would you define war? What is missing? What should be removed?

Study Guide for Chapter 7

Lambs and Lions

Benchmark: What are your thoughts concerning the wrath of God? How does that correlate with what you know about Jesus?

1. Read Exodus 34:6-7. Describe the *Divine Paradox* of this text. How does the passage present a conundrum?

2. From Chapter 7, what is every man's stance before God as a function of their sin? What does justice require? What passages back this up?

3. Consider the phrase, "God hates the sin, but loves the sinner." What are your thoughts concerning this phrase? Why?

4. What surprises you in reading about the wrath of God in this chapter? Why? Do you feel that it is fair? Why or why not?

5. From Chapter 4, where do we see the love of God in the Old Testament? For whom does God, as describe in the Old Testament, have a special affinity?

6. Why do most people focus upon the loving aspects of God's character as described in 1 John 4:8 or John 3:16? What is the danger in this regard?

7. Read John 14:6. What does Jesus mean? How might this verse offend or confront many? Why?

8. From Chapter 7, where in Scripture do we see some different aspects of Jesus' character?

9. Why do most people—and many churches!—neglect this aspect of Jesus' character?

10. Read Colossians 1:17. What does this verse mean? What are the implications of this verse concerning warfare?

11. Return to the *Divine Paradox* of Exodus 34. Can you better answer the question about this verse now? How does God satisfy this paradox?

Study Guide for Chapter 8

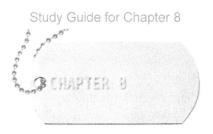

Centurions and Soldiers

Benchmark: What is the soldier's standing before God? How does God view the soldier?

1. From Chapter 8, consider the descriptions of the Roman soldier's part in the crucifixion of Jesus? How active of a role did they play? How do you think God would view their actions? Why?

2. Read Acts 2:23 and Acts 4:27-28. According to these verses, why was Jesus crucified?

3. As Jesus hung dying upon the cross, who was the first to recognize Him for who he was? What did he say?

4. Who was the first Gentile convert? How does Luke describe him?

5. Luke describes the encounter between John the Baptist and several soldiers. John tells them, "Do not extort money from anyone by threats or by false accusation, and be content with your wages." What are the implications of this verse? Restate John's directions in your own words? What does John not tell them?

6. Peter tells the man, "God shows no partiality." What are the implications of this verse for the soldier and his standing before God?

7. From Chapter 8, what four statements may we make concerning God's regard for the soldier? Do you agree with them? Why or why not?

8. What are the two competing but dangerous views concerning the soldier's stance before God?

9. Knowing what you've learned, does this change your view concerning the soldier's relationship with God?

Study Guide for Chapter 9

Saved Rounds

Review

1. What are the two purposes given that God left the nations? Why?

2. From Chapter 2, define the historical-grammatical method of biblical interpretation. Why is interpretive method important?

3. Rewrite the following phrase in your own words, "Orthodoxy always precipitates orthopraxy."

4. In your opinion, can we trust the Bible? Why or why not? Review what Chapter 3 says concerning the matter.

5. Read Romans 8:28. What are the implications of this verse?

6. From Chapter 3, define the sovereignty of God, providence, and concurrence.

7. What was the wickedest act in history? Who directed and executed this act?

8. Read Deuteronomy 29:29. What are the implications of this verse?

9. Read Genesis 1:26 and 2:7. How is man distinct from the rest of creation?

10. From Chapter 5, what is God's initial command to Abraham? What are the 3 aspects of God's covenant with him?

11. Read Judges 3:1. From this verse, who left the nations in Canaan? For what purpose?

12. Read Exodus 34:6-7. Describe the *Divine Paradox* of this text. How does the passage present a conundrum?

13. Peter tells the man, "God shows no partiality." What are the implications of this verse for the soldier and his standing before God?

Study Guide for

Section 2:
War and the Mind

Study Guide for Chapter 10

The Utility of War

Benchmark: Define evil in your own terms. Have you ever been confronted with something you truly consider evil?

1. In review, what is the previously discussed reason that God left Canaan in the land? What is the second reason we'll examine?

2. Read Psalm 144:1-2. What are the implications of this verse pertaining to warfare?

3. From Chapter 10, what are your thoughts on the Rape of Nanking? Is it even necessary to discuss? Why or why not? What shocks you the most about the event?

4. Consider these two statements from Chapter 10. What are your thoughts on these two statements? Are they true? Why or why not?

a. All men possess the capacity to rape and kill and at some level, the desires.

b. Organization and collaboration amplifies the effects of raping and killing.

5. Chapter 4 states, "The presence of evil in this world necessitates resistance, fighting this evil, overcoming." Do you agree with this statement? Why or why not?

6. In a world of subjectivity, how may one distinguish between what is good and what is evil?

7. From Chapter 4, what are three places where God calls a believer to confront evil?

8. If you've fought consider the war. If you haven't, then consider a contemporary war? What were the stated national objectives? Were they just? Were there other potentially less just objectives? Was injustice committed at any level? What is your overall view of the conflict?

9. What do you think about God teaching through war? What would be the primary lesson we might learn by participating in warfare?

Study Guide for Chapter 11

The Mind of the Warrior

Benchmark: Meditate and list the major events that have shaped your life, how you think, what has happened?

1. Do you consider Christianity to be a thinking man's religion? Why or why not?

2. Read Romans 12:2. How is a man transformed? What is the outcome of that transformation?

3. What are your thoughts concerning the seeming conflict between science and faith, between reason and Christianity? Why?

4. What is the most traumatic event you've experienced? Why? Have you ever feared for your life as SSG Conway did?

5. In what ways does SSG Conway's story resonate with you? Can you identify with his situation? What struggles do you share with him?

6. Consider Hannah, SSG Conway's wife? What do you think was her primary struggle? How could she have dealt with this? Have you ever experienced anything similar or know someone who has? How did you(they) deal with it?

Study Guide for Chapter 12

The Struggle of War

Benchmark: What are your thoughts concerning the value of man? How does this dictate your thoughts and actions? Does it? Why or why not?

1. Read Romans 1:25. What did men do that so angered God?

2. Read Romans 1:22-23. What did men claim? Where do we see that truth today?

3. Contrast a biblical mindset from an evolutionary mindset. Give some specific examples of how they might differ.

4. Read Acts 10:34. What does this verse say concerning men? What are the implications?

5. Restate the three supporting points to Christ being the great equalizer.

6. From Genesis 1:26, how are men different from all other created things? What does this mean? How does it matter?

7. What are some ways we see the image of God in man?

8. What are the two greatest commandments?

9. If man is truly the pinnacle of creation, what are the implications concerning daily life? Warfare?

10. Meditate on your own thoughts concerning the value of a man. Are there those you devalue or hold a lesser view of? Are there groups you view as inferior? Why?

Study Guide for Chapter 13

Killing the Mind

Benchmark: In your opinion, what is the greatest trauma soldiers suffer in general?

1. From Chapter 13, what are some of Grossman's observations? Why do you think this is?

2. What is your opinion concerning the righteousness or potential righteousness of killing? Why?

3. What is the dichotomy spoken of in Chapter 13?

4. What is the primary, root cause of trauma?

5. Define "trauma" in your own words?

6. Give an example, personal or otherwise, of a traumatic situation you've seen. What made it traumatic?

7. Define moral injury.

8. What distinguishes moral injury from trauma?

9. Give a specific example of a moral injury you're privy to.

10. What is it that drives the disorder label associated with trauma?

11. List some trauma mitigating factors? What is it about these factors that make them effective?

12. Read the quotation at the top of page 189. Do you agree or disagree? Why or why not?

Study Guide for Chapter 14

Resolution

Benchmark: In your opinion, what is the greatest struggle most soldiers face today?

1. What three factors shape the response to trauma?

2. Have you ever been a part of what you considered a brotherhood? What was it? What did that mean to you? How did it feel? Are you still a part of it? Why or why not?

3. Read Proverbs 3:5. Why does the writer make this assertion?

4. Read Philippians 4:8. With what should a believer fill his mind? Where would he find these things? Relate this to the concept of mediation.

5. What was Conway's poignant question? Have you ever been confronted with a situation that forced you to ask the same question?
6. Describe the incomplete reconciliation from Chapter 14.

7. What role should the local church play in ministering to the warrior?

8. What are your opinions concerning popular treatment options for those suffering from traumatic stress disorders such as psychotropic drugs, homeopathic treatments, eastern meditation (yoga), or secular counseling?

9. What are your thoughts on the statement, "The real sacrifice of the soldier is the acknowledgment of this burden and a willingness to bear it on behalf of those we love."? True or not? Why?

Study Guide for

Section 3: War and the Heart

Study Guide for Chapter 15

Hearts on Display

Benchmark: How should a soldier conduct himself? In what fashion should a Christ-follower conduct himself during the course of war?

1. In your own words, what is the warrior ethos? From where does the warrior ethos come?

2. From Chapter 15, what was the other crew member's attitude towards Muslims? From where does an attitude like this derive?

3. In the conduct of war, does intent matter? If so, how?

4. If you've participated in combat, do you have a 'switch' as described in Chapter 15? If not, why? If so, why is the switch necessary?

5. Read the *Warrior Ethos* on page 213. Is this useful? What qualities does it highlight? How does it square with your definition?

6. What were your thoughts reading the spin-up of Weeks and Worrell? Have you ever been called upon in such a fashion? How did you respond? What drives a man's response in a situation like this?

Study Guide for Chapter 16

The Theoretical Heart

Benchmark: Consider your own heart. How would you describe the condition of your heart?

1. Define leadership in your own words. What do you think of Eisenhower's definition on page 218?

2. Define the 'heart' from Chapter 16.

3. Biblically speaking, what is the function of the heart?

4. In your opinion, is man inherently good or bad?

5. What do Romans 5:12, Ecclesiastes 9:3, and Jeremiah 17:9 say about the heart of a man? Do you believe this?

6. Define total depravity per Chapter 16.

7. What two points concerning 'goodness' does Chapter 16 make? Why are they important?

8. What is the fundamental difference between Christianity and every other religion? Why is this important?

9. Define 'regeneration'.

10. Where is the primary work of God in a man?

11. For what reason does God change a man's heart?

12. What is the difference between biblical, godly love and worldly love?

13. How can this relate to warfare?

Study Guide for Chapter 17

The Heart in Practice

Benchmark: What ought to be the primary business of a man? Who gives us the greatest example of this?

1. Read 1 Corinthians 13:11. What are your thoughts about this verse? Are these ideas widespread today?

2. Read John 15:13. How might this apply to warfare? How might this apply to a man's life outside of warfare?

3. Read John 15:12. What is Jesus' command? How should a man love? What does this mean?

4. Contrast a biblical view of war and its motivations with those spelled out by Sun-Tzu and Clausewitz?

5. What are your thoughts concerning the reconciliation of love with wartime motivations? Is this valid? Useful?

6. What are your thoughts on the Bearded One's exhortation?

7. From where does the spirit of atrocity come?

8. What factors check sinful behavior during war?

9. What two things amplify both goodness and sin?

10. Should the warrior have a 'switch'?

11. Have you known of a sacrifice such as the one made by Worrell and Weeks? Does a man need a physical battlefield to sacrifice in such a way? Why or why not?

Study Guide for Chapter 18

Heart of a Leader

Benchmark: In your opinion, what is the most important aspect of leadership?

1. What happened at the Last Supper as pointed out on page 245? Had you heard about this previously? Does this surprise you?

2. What is Jesus' response to the disciples in this instance?

3. How does Jesus' response square with the things that He taught?

4. How does ADP 6-22 define leadership? Is this valid? Is anything missing in your opinion?

5. Think of the good leaders you've worked for before, in the military or otherwise. What are some of their characteristics? Are there common characteristics to these good leaders?

6. Conversely, consider the bad leaders you've worked for before. What are some of their characteristics? Do they share any common traits?

7. Have you worked for an inspirational leader? What made this man or woman inspirational?

8. What are the three characteristics of a godly leader per Chapter 18? Do you agree with these three?

9. How does humility contribute to leadership?

10. What are some dangers for a leader who is not humble?

11. How does stewardship relate to leadership?

12. How does service relate to leadership?

Study Guide for Chapter 19

Consecrated Hearts

Benchmark: What is the toughest battle that lay before you? What are you doing to prepare for it? What should you be doing?

1. Read Joshua 3:5. What is Joshua's command to his men?

2. Why did he issue this command? What were they preparing to do?

3. What does it mean to "consecrate' yourselves?

4. In what ways do you see the daily struggle of life as discussed in Chapter 19?

5. How would you rate yourself in how well you consecrate yourself daily to prepare for the difficulties in your life?

6. Think of a time you've been tested in life, pushed to the limit of your physical, emotional, mental, or spiritual capacity. Were you ready? Had you consecrated yourself? Did it matter?

7. Consider Warrant Officer Thompson's situation. What challenges/dilemmas did he face that day?

8. Review the Benchmark question from Chapter 14. What is the condition of your heart? Does it need to be changed? Be honest.

9. What might be keeping this from happening?

Study Guide for

Section 4:
War and the Soul

Study Guide for Chapter 20

The Soul and the Law

Benchmark: Name from memory as many of the Ten Commandments as you can. Why did God give the Ten Commandments?

1. "All things possess a spiritual component." Do you agree with this statement? Why or why not?

2. What steps have you seen the military take to imbue the warrior spirit in its soldiers, whether you've personally served or not?

3. Is it effective? Why or why not?

4. In review, where does a man's natural resistance to killing his fellow man originate?

5. Read Galatians 3:24. What does it mean that the law (Ten Commandments) is our 'guardian'? What other words to describe the law could be used?

6. Read Romans 7:7. From this verse, how is the law a schoolmaster or a tutor?

7. Why do we need the law to serve in this capacity?

8. Read James 2:10. What does this verse tell us?

9. What might generate guilt and shame in some warriors?

10. What is the most accurate rendering of the sixth commandment? Why is this distinction necessary?

11. Why do you believe the Ten Commandments are not widely known in contemporary society? What could be the impact?

Study Guide for Chapter 21

The Soul of a Warrior

Benchmark: What do you know of Kind David? What did he do? What was his relationship with God? What role did he play in ancient Israel? What happened to him?

1. Chapter 21 speaks of the feminization of the west, writ large. Do you agree with this assessment? Why? Where do you see this manifest? Is it important?

2. In what two ways do men err in response? What are your thoughts on this?

3. What is a primary title of David? From where did this title come?

4. Who are the two groups of warriors? What did they do? Consider them in light of the intentionality of Scripture, that all Scripture is inspired by God. Why would God reveal the existence of these warriors?

5. What is a major attribute of these men?

6. How did David rise to power? By what means?

7. What were David's dying words to his son Solomon?

8. From what may we infer about verses like 1 Samuel 18:17 and Psalm 89:19-23?

9. Describe David's transgression. What were the major aspects of his sin, both obvious and maybe not as obvious?

10. Consider David the worshipper. How do you reconcile this role with that of the warrior?

11. Based upon what you now know about David, how does or should this encourage the walk of the warrior?

Study Guide for Chapter 22

War and the Soul

Benchmark: In your words, what is the Gospel?

1. What would you consider a 'good death'? Why?

2. Read Matthew 6:27 and Psalm 29:5. What do these verses remind us? What then must we do?

3. Read Psalm 37:23. What is the significance of this verse? How would this verse affect our actions?

4. From 1 Corinthians 15:55, what is the universal scourge of man? What is the solution to this problem?

5. Consider Romans 14:10 and Romans 15:15. How should these verses affect what we do? How we live?

6. Why must God hold man accountable for his sin?

7. What is the fate of all who will not believe on Lord Jesus?

8. Read Matthew 7:21-23. What is the application of this verse? Why do you think this is?

9. What has been the most important things in your life? Have they changed over the years? Why? What matters most to you currently?

10. Have you ever felt that God could/would not forgive you of a particular sin? Why? What does 1 John 1:9 tell us about this?

11. What scars do you bear today? Are they scars of combat? Other?

12. Would you know Him today? Would you submit to Him today? If not, why?

13. What caveats does God offer the warrior? In light of this, return to question 12?

14. What is keeping you from repenting of your sin and submitting to Jesus today?

Study Guide for

Section 5: War and the Nations

Study Guide for Chapter 23

God and the Nations

Benchmark: What is your opinion of your current government? Is it godly? What do you see as your role and/or duty to the government?

1. Define *transcendence* and *immanence*. How does this apply in God's relationship to governments and/or nations?

2. Where do you see God working on a national or global level? What are the impacts of this?

3. Read Proverbs 21:1. What are the implications of this verse? How does it display *concurrence*?

4. Who deployed Assyria against Israel? Did Assyria seek to be an instrument of God's wrath? What are the implications of this?

5. What happened in A.D. 313? Why is this significant?

6. What does the Bible say concerning the Christian and the government? What biblical text(s) supports this?

7. Define *sacralism*. Why is this important? Where do we see *sacralism* today?

8. Who ordained governments? For what purposes?

9. Compare and contrast earthly kingdoms with the Kingdom of God.

10. Why is this important? In regards to kingdoms, what is God's primary business?

11. Why might this be confusing for the warrior?

12. When would it be biblically justified to resist the government? What about armed resistance?

13. What are your thoughts on the section concerning Islam? How should the world deal with Islam?

Study Guide Conclusion

Review

Benchmark #1: What do desire to get out of this study? What are your objectives? Why? We'll examine these at a later date?

Benchmark #2: How would you consider your faith, currently? In what do you place your faith? Do you have any faith? If you are a Christian, how actively do you practice your faith?

Benchmark #3: What are your thoughts about or impressions of God? His character? His attributes? His work? From where do you get your ideas concerning God?

Benchmark #4: Review your benchmark answers from last week. Can you identify anything in your life that contributed to your ideas concerning God?

Benchmark #5: What is the origin of war?

Benchmark #6: What is your understanding of the overall theme of the Bible? Is there one?

Benchmark #7: Describe a time you've been afraid in battle. If not in battle, a time that you've been fearful. What generated the fear? What did you do? How did you deal with the fear?

Benchmark #8: What are your thoughts concerning the wrath of God? How does that correlate with what you know about Jesus?

Benchmark #9: What is the soldier's standing before God? How does God view the soldier?

Benchmark #10: Define evil in your own terms. Have you ever been confronted with something you truly consider evil?

Benchmark #11: Meditate and list the major events that have shaped your life, how you think, what has happened?

Benchmark #12: What are your thoughts concerning the value of man? How does this dictate your thoughts and actions? Does it? Why or why not?

Benchmark #13: In your opinion, what is the greatest trauma soldiers suffer in general?

Benchmark #14: How should a soldier conduct himself? In what fashion should a Christ-follower conduct himself during the course of war?

Benchmark #15: Consider your own heart. How would you describe the condition of your heart?

Benchmark #16: What ought to be the primary business of a man? Who gives us the greatest example of this?

Benchmark #17: In your opinion, what is the most important aspect of leadership?

Benchmark #18: What is the toughest battle that lay before you? What are you doing to prepare for it? What should you be?

Benchmark #19: Name from memory as many of the Ten Commandments as you can. Why did God give the Ten Commandments?

Benchmark #20: What do you know of Kind David? What did he do? What was his relationship with God? What role did he play in ancient Israel? What happened to him?

Benchmark #21: In your words, what is the Gospel? _Do you believe it? Why or why not?_

Final Benchmark: How has this study changed you? Benefited you? Has it? What do you still have questions about? Do you have a plan to pursue the answers?

Personal Note

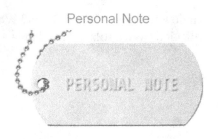

Brave Rifle,

I am praying for you, time now. I'm praying that you'll see the Gospel for what it is. I'm praying that you'll understand that you are a sinner in need of a Savior and that savior is the Lord, Jesus Christ. Repent of your sin and turn to faith in Christ. Confess with your mouth that Jesus is Lord and believe in your heart that God raised Him from the dead.

He will change your life. I promise you this.

Looking back, I can scarcely remember my life before Christ. My only regrets in life involve wasting so much of my life pursuing things other than Him.

If you are already of Christ, I pray that this work has been a blessing to you. I pray that you've come to an understanding of the theology of war and that perhaps this reconciliation has helped to heal some scars upon your very soul. Again, turn to the Lord Jesus. He alone is the healer. He makes all things new. It's what He does. Of this, I assure you.

Brad

www.the413project.com

Mission: Common People Empowering and Serving Others to Accomplish an Uncommon Good.

Following Pentecost, Peter, John, and the others preached with newfound passion and authority. They boldly and fearlessly proclaimed Christ. The religious authorities observed,

> *Now when they saw the boldness of Peter and Joy, and perceived they were uneducated common men, they were astonished. And they recognized that they had been with Jesus.*
> *Acts 4:13*

As a result, the early church exploded across the known world. The 413 Project seeks to empower the local believer to accomplish that to which God is calling them. Our current projects include this work and other associated publications, the 413 Report, the Deep Dive Bible Study Series (the Brave Rifles Study Guide is the first in this series), and a local clothing closet focusing on foster kids coming into the system. Future projects include The Colorful Family (a children's book series highlighting multi-racial adoptive and foster families), more books from Bradford Smith, and a podcast to accompany the 413 Report.

Current Contibutors

Bradford Smith—author, contributor
Ami Smith—future author, contributor, clothing closet co-director
Jennifer Drake—clothing closet director
Jayson Rivas—Administrative and Technical Advisor, JHR Photography
Megan Gentleman—editing

About the Author

BRADFORD SMITH has been married to his best friend Ami since 2001 and they stay busy raising their nine children together. A West Point graduate, Brad has served on active duty since 1995 including multiple combat deployments to Iraq and Afghanistan.

In 2007, Brad surrendered to the call to preach and I 2011, he joined the staff of The Way, a Baptist church in Clarksville, as the Missions Pastor where he continues to server bi-vocationally.

In 2010, Brad and Ami opened the Clarksville Covenant House for teenagers who age out of the foster system and in 2013, Brad graduate from Liberty Theological Seminary with a Masters of Divinity. Once the Army releases him from active duty in 2018, he plans to pastor full time, continue writing, and run the Covenant House alongside his wife.

www.thewayofclarksville.com
www.thewayofclarksville.com/covenant-house

CPSIA information can be obtained
at www.ICGtesting.com
Printed in the USA
BVHW031150150322
631528BV00011B/27